Carnivorous Plants

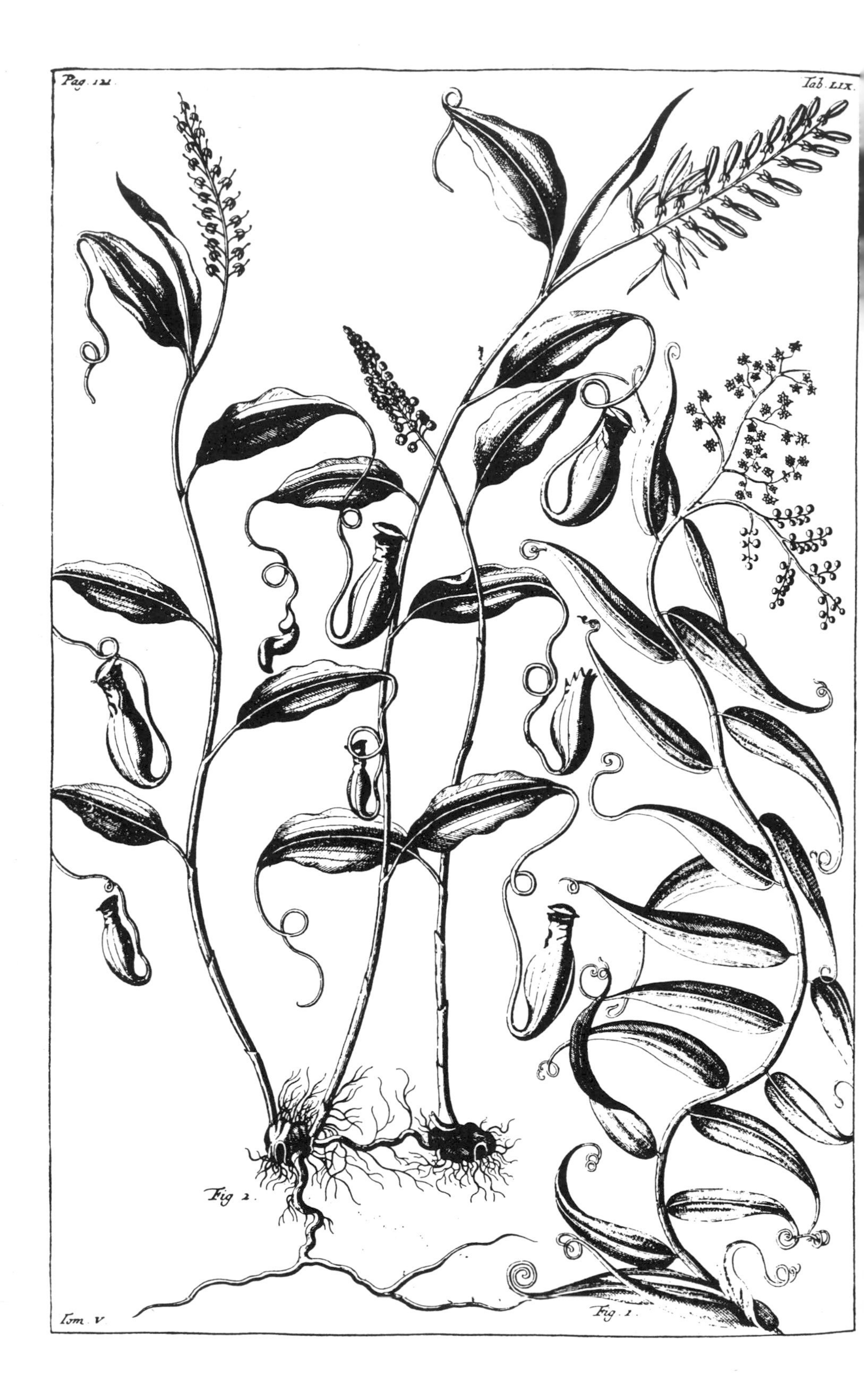

Pag. 121
Tab. LIX.
Fig. 2.
Fig. 1.
Tom. V

Carnivorous Plants

Randall Schwartz

Edited by Deborah Leavy

Praeger Publishers
New York · Washington

FOR MY PARENTS

and for Dr. Fred Norris,
Professor, Department of Botany,
University of Tennessee, Knoxville

Frontispiece: The earliest known illustration of *Nepenthes (Nepenthes mirabilis)* from Rumphius, *Herbarium Ambionense,* published in 1747. The plant on the right is *Flagellaria indica.*

Published in the United States of America in 1974
by Praeger Publishers, Inc.
111 Fourth Avenue, New York, N.Y. 10003

Library of Congress Cataloging in Publication Data

Schwartz, Randall.
Carnivorous plants.

Bibliography: p.
1. Insectivorous plants. 2. House plants.
I. Title.
SB432.7.S36 635.9'3 73-19438
ISBN 0-275-51580-X

Printed in the United States of America

Contents

Preface

Science-fiction movies aside, there are no man-eating plants. Yet carnivorous plants are a fascinating quirk of nature, trapping their victims with methods that would put a science fiction writer's imagination to shame. Of the more than 450 species of carnivores, some are small fungi that grab wandering amoebae, while others are capable of devouring lizards, rabbits, and giant tropical rats.

I first became involved with these strange plants when I purchased some three-for-a-dollar Venus flytrap bulbs when I was twelve. They grew, and my friends and I were fascinated by them. A little research turned up other carnivorous plants that grew locally, and soon I was searching nearby fields for sundews and butterworts. There's a certain thrill of satisfaction about finding them in the wild.

Carnivorous plants have always been a part of my life. No matter where I've lived, whether I had a small apartment or a greenhouse, I've always managed to grow them.

I've found carnivorous plants quite easy to live with. They're not much more demanding than ordinary house plants, but much more fun. After all, you can't really play with a palm tree or an African violet. Carnivores are practically pets!

More seriously, carnivorous plants can be an organic alternative to insecticides. There are plants so efficient that they capture hundreds of ants each day, and others that consume literally thousands of mosquito larvae. No flies, ants, or roaches around my place; not even a gnat.

Charles Darwin, Sir John Burdon-Sanderson, and other great naturalists have studied and written about these plants. While some facts have been ascertained, many aspects of their behavior still remain unexplained. Nerve activity has been discovered which resembles that of animals; yet no one seems positive of how some carnivores move.

The purpose of my book is to popularize these amazing plants. I've talked about where to see them, where to find them, and how to grow them. I've provided accurate information and illustrations of more than forty species of carnivores. And because the plants themselves are so much fun, I hope to amuse and entertain you as well. All this, to lure you into the incredible world of carnivorous plants.

Acknowledgments

Many people have shared my enthusiasm for carnivores, and helped by contributing their time and effort to my pet project. Among them: Peter Max, whose famous sense of graphics added to the book's design. Artist/designer Elaine Grove's input was invaluable, both in design and illustration. Susan Lazarus processed and printed my photographs.

For his help with the plants and content, I must thank Joseph Mazrimas.

Mary Northrop offered assistance continually throughout this project. She and her late husband, Bill Northrop, have been instrumental in popularizing these plants for the past decades. The Berkeley Botanical Gardens, Brooklyn Botanic Gardens, and New York Botanical Gardens all provided research facilities.

Super-Snipe Ed Sommers and his fabulous memory helped out with the old comic book illustrations, as did the staff of *Monster Times.*

And Barbara Weiner and Calo Rios deserve many thanks for always being there when things were needed.

For permission to reproduce cartoons, I would like to thank the following: Eerie Publications, New York, reproduction on page 117; National Periodical Publications, Inc., New York, reproduction on page 122; and *Mad Magazine* (© 1958 by E. C. Publications, Inc.), reproduction on page 123. In addition, I would like to thank the Field Museum of Natural History, Chicago, for allowing me to use excerpts and to reproduce the engravings on pages 119 and 121. With the exception of the excerpt from *The Patchwork Girl of Oz* (© 1913 by L. Frank Baum) on page 121, all excerpts in Chapter 6 are reprinted from the Field Museum of Natural History's Botany Leaflet 23, *Carnivorous Plants and "The Man-Eating Tree"* by Sophia Pryor (© 1939, Chicago).

I would like to thank Harvey Dickler for providing the photographs on pages 73, 94, 97, and 101, and Joseph Mazrimas for providing the photograph on page 51.

Introduction

Carnivorous plants are a mystery to most people. The most basic characteristic of these exotic plant species, and the source of our fascination with them, is their method of survival, unique among plants. They all lure and trap animals, and get nutrients from their digested prey. Growing generally in nitrogen-poor soil, the plants evolved their carnivorous trait to stay alive in an environment that wouldn't support most other plants. Since their victims supplied them with the extra nitrogen they needed, carnivorous plants did not have to compete for the limited supply of nutrients available from the soil and water within their domain.

To capture their prey, the plants have developed three different trap styles. The most familiar is the active trap, typified by the fast-action Venus flytrap. Its reflex movement makes it the most animal-like. The semi-active trap utilizes movement, merely as a supplement to the adhesive that captures the prey. Sundews and butterworts are in this category. As for the passive traps, they rely totally on ingenious design to ensnare their victims. Traps like the pitcher plants are virtually obstacle courses and the plants are rewarded with a steady diet of insects.

Unfortunately, another characteristic of carnivorous plants is their growing scarcity. Boglands, the natural habitat of most carnivores, are making way for housing developments. Searching for plants in the wild, in areas called natural stations, is becoming increasingly difficult. While I've indicated the areas where each plant grows, I've decided not to point out exact locations, fearing that would invite a stampede. Most of the easy-to-grow species have been cultivated in nurseries, so when you buy them you're not depleting the natural stock.

Hunting for carnivores in the field does have a special excitement. But if your search is successful, treat the plants as endangered species. Be careful to take only what you can care for. Please don't be greedy, unless you're saving the plants from an encroaching bulldozer!

Just in case your last course in botany is only a distant memory, here's a quick reminder about plant names. Since plants grow around the world, labeling them has been standardized in Latin to avoid the confusions of translation. It may look complicated, but the method is simple. Each plant has two names: first, the genus, describing a fairly large class of plants; and second, the species, describing the specific type. Occasionally there are subspecies, but if you remember the system they're easy to follow.

Carnivores often show a tremendous variety within a particular species, which affects the shape, pigmentation, size, or other details. I've tried to use average plants from each species, but your own plants may vary somewhat.

Although many carnivorous plants have beautiful blossoms, I have generally shown the plants without their flowers in order to better show the details of the traps. Contrary to science fiction, the flowers do not trap, but they are a pleasant bonus to the carnivores.

Carnivorous Plants
MAX

1. Active Traps

Active traps—the very words evoke images of vicious plants devouring their victims. And the image is not inaccurate.

Active plants display rapid motion in the capture of their prey. Two kinds of traps are commonly found within this group of plants: fast-closing miniature bear traps or elaborate fast-moving trap doors with delicate triggers awaiting that fatal touch.

The active carnivorous plants in this chapter can be grown at home. Other active carnivores are described in Chapter 4.

Venus flytrap – *Dionaea muscipula*

Don't look for the Venus flytrap in the tropical jungles of Africa or South America. Its only native habitat is the area around Wilmington, North Carolina, where it grows in what is thought to be an ancient meteor crater now filled in by a bog.

Actually, the name *flytrap* is a misnomer. Although it will definitely trap flies, the plant's primary food source is large ants, and it has even been known to catch brine shrimp during rare salt water flooding.

Looking like a cluster of bear traps, the Venus flytrap lies in wait for its victims. Insects are attracted by a sweet nectar inside the trap. There, on each side of the trap, are three dark trigger hairs, and they are extremely sensitive.

When an insect touches a trigger hair once, nothing apparently happens. The plant has been alerted, but the hair mechanism could have been activated by the wind, or a piece of dust. When the prey brushes against a trigger hair again, that second signal causes a reaction; nerve-like electrical impulses pass through the plant, and the trap closes instantly. The "teeth," called cilia, interlock and form a cage, enclosing the victim.

Small insects can easily crawl out between the cilia. This is the plant's method of throwing back the small fish, since it would take more energy to digest a tiny insect than could be gained from consuming it. But larger insects struggle to get out, repeatedly touching the trigger hairs. These signals cause the trap to close completely, totally sealing in the prey. The trap fills with liquid and the victim drowns.

The liquid, a digestive enzyme, breaks down the soft parts of the insect so it can be absorbed by the plant. This process takes several days. When the plant has digested all of the insect it can use, the trap reopens, exposing the insect's skeletal shell, which is soon swept away by wind or rain. If the trap was mechanically triggered and captured no prey, it reopens within a day or so.

Apparently the closing of the trap is caused by the upsetting of tension in the outer epidermal cells of the trap. When the trap is open, these cells are in tension balanced against an inner layer

known as parenchyma cells. When an insect enters the trap, electrical signals from the trigger hairs release an enzyme which causes the outer epiderm to stretch. This stretching causes the wall to warp shut, all within a fraction of a second. Further stretching forces the trap to seal its victim within. There is marked protoplasmic activity in the individual cells. Conversely, the trap reopens with the increased growth of the inside walls. The outer wall then grows slightly to reset the trap. Because of the amount of growth involved, each individual trap is usually capable of only three or four closings in its life span. Then it turns black and drops off, and the plant produces new traps.

A longtime favorite as a greenhouse curiosity, the Venus flytrap is easy to grow. Give it strong sun and high humidity. Sphagnum moss, straight or mixed with a little sand and kept wet, most closely resembles its natural bog habitat.

The plants are almost foolproof, but if the traps fail to develop, then the humidity is probably too low. If the sunlight is poor, the traps will be smaller, leaves longer. Look for a reddish tinge inside the traps – it's a sign the plant is getting enough sun. Although some advise against it, I've had success fertilizing my flytraps lightly with a very weak solution poured over the leaves monthly during the growing season. Be careful to use organic fertilizer such as fish emulsion and periodically run copious amounts of water through the pot to prevent any possible fertilizer build-up.

The plant forms a rosette of traps, which grow in either a flat or somewhat upright position, depending upon the amount of sunlight available. Usually the whole plant is only a few inches across, with individual traps about an inch long. When found naturally, several plants form a tangle of traps, so that every wandering insect is bound to trigger one of them.

Like most plants, the Venus flytrap produces little growth in winter and will in fact survive frost. It blossoms in the spring with a cluster of white flowers. Through spring and into summer, it

Close-up of *Dionaea muscipula.* Note small, sensitive trigger hairs.

grows very quickly, producing one trap per week and maintaining about six traps per plant.

Flytraps are available commercially as bulbs, and occasionally as young plants. As the plant grows, the bulb expands in layers. Inspect the bulb when you're repotting. If it's large enough, you can split off a section of the bulb and have two plants.

Flytraps can also be reproduced by seed; just sow them on peat and keep them moist. Or reproduce them vegetatively with leaf cuttings. Gently peel off a leaf in its entirety, being sure to get the white part near the base. Lay the leaf on damp moss, tucking the base of the stem partway into the moss. Keep it moist but not wet and it should root in a few weeks. Treat it as a normal plant after its third new leaf.

Dionaea muscipula in blossom.

Bladderwort – *Utricularia*

Imagine a round bubble with a closed door, a feather sticking out below the door. Something comes by and touches the feather. The door flies open, the victim is sucked into the bubble, and the door closes. This is how the trap of the bladderwort works, and its astonishing mechanical quality is unequaled in the botanical world.

Most species of *Utricularia* are aquatic and grow floating in water. Some others are terrestrial, and a few are epiphytic, growing in moss. One species grows in the pool of water found at the base of the leaves on large bromeliads (plants in the pineapple family). Some others have blossoms that are as showy as the finest orchids.

But it is the trap of the bladderwort that makes it unique. Its mechanical precision, amazing in itself, is even more amazing when you consider its size. The largest is only five millimeters (one fifth of an inch), the smallest 0.3 millimeters (one hundredth of an inch).

Let's examine the inner workings of the bladderwort. The bladder, or trap, is an oval balloon with a double-sealed, airtight door on one end. When the door is closed, the bladder expels water through its walls, creating a partial vacuum inside. This negative pressure causes the walls of the trap to collapse slightly inward when the pressure is at its peak.

Jutting out near the door is the trigger. Sometimes forked or branched, sometimes single, the device is always deadly. The instant an unsuspecting prey touches the trigger, the door opens. The vacuum inside causes an immediate suction and the victim is gulped up by the plant.

At the bottom of the trap, a small pool of water contains digestive enzymes that disintegrate the victim. The bladder expels any excess water that may have been sucked in, creating another partial vacuum and thus resetting the trap.

A greatly magnified view of a branch of *Utricularia*. Note upper trap has captured prey.

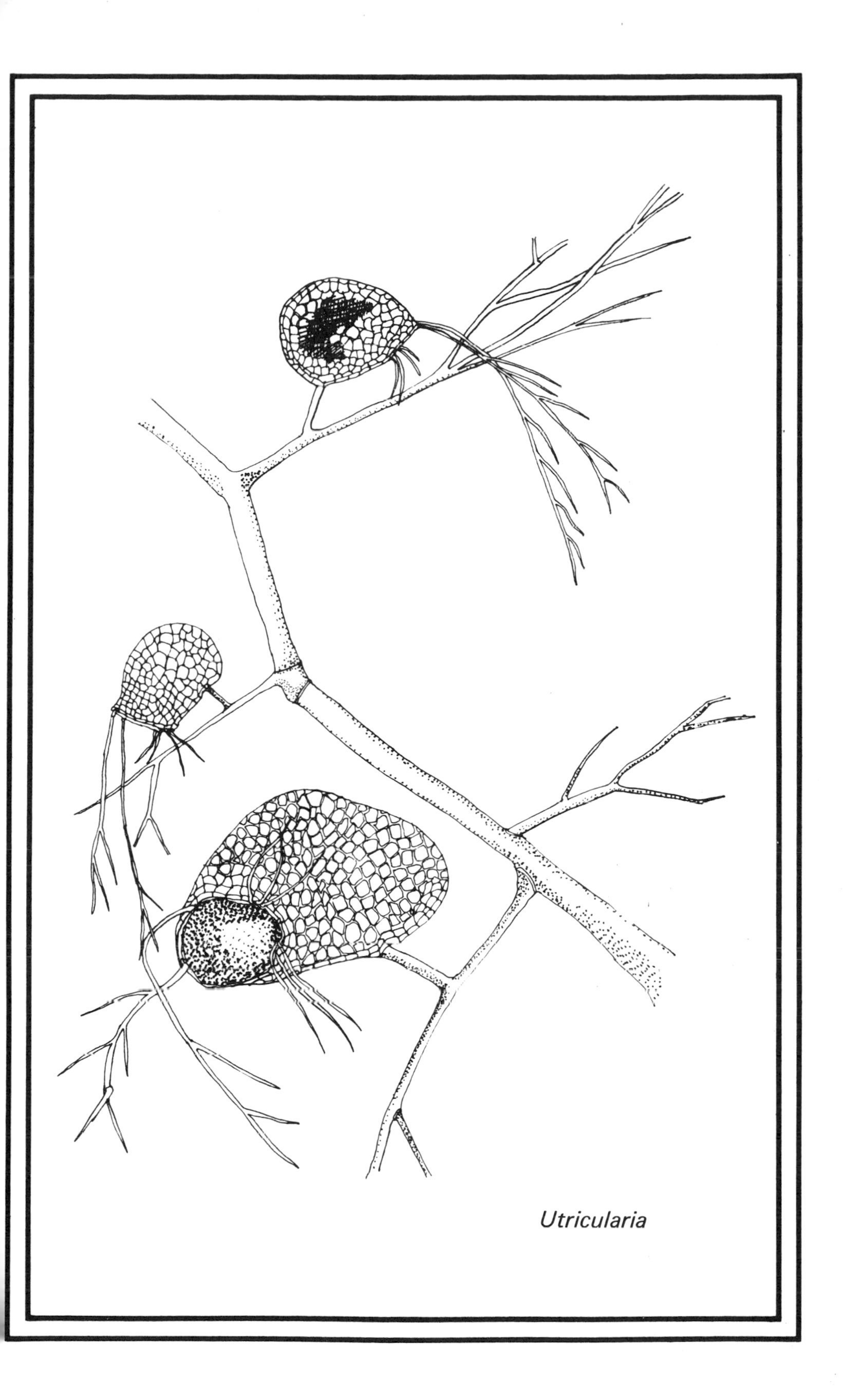

Utricularia

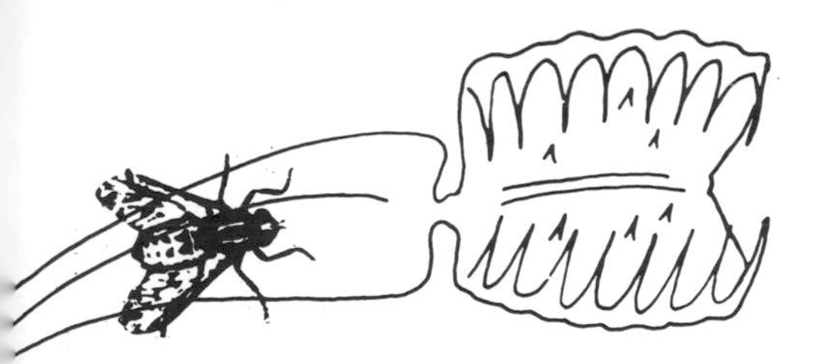

Most bladderworts eat water fleas and mosquito larvae, and occasionally a tiny tadpole. To digest bigger than bite-size victims, the traps make successive gulps, holding the prey tightly between gulps until the trap is reset.

It would be interesting to see the plants cultivated in stagnant water where mosquitoes breed. The hungry traps might be an organic solution to that problem; they certainly have the potential.

For your own home growing, you'll have to keep the aquatic species outdoors; they don't lend themselves to indoor cultivation. You don't need a pond, though – a large plastic bucket will do. Just keep the water level up, with a constantly dripping flow, if possible.

The species that don't grow in water may be easier for you to grow. The epiphytics do well in a mixture of sphagnum moss and osmunda fiber, or just plain sphagnum, kept moist at all times. The terrestrials prefer acid soil, either alone or mixed with sand or sphagnum. If you're unsure of the kind of soil you have, use live sphagnum.

2. Semi-active Traps

Tentacles tipped with a lethal, sticky fluid are ready to clutch their victim. But this is a two-stage trap. The prey must already be caught in the trap's adhesive before the movement begins. The insect's struggle triggers the slowly tightening grip.

Not all the semi-active plants have tentacles. Some have wide, flat, sticky leaves. When an insect is caught, the leaves roll upward to cup digestive fluid, and to surround the prey with more of the gluey surface.

In this chapter are many of the semi-active plants that can be grown at home. More difficult plants can be found in Chapter 4.

Sundew – *Drosera*

The sight of a delicate *Drosera* sparkling in the sunlight is deceptive, belying the plant's deadly nature. Perhaps the most beautiful of all the carnivores, its Latin name is derived from the Greek word for dewy. But the sundew glistens because every leaf is covered with drops of sticky fluid, carried on the ends of tiny tentacles.

When an insect lands on a leaf of the sundew, attracted by its color and aroma, the fatal struggle begins. Trying to escape from the gluey tentacles, the victim stimulates the production of more fluid, in which it wallows, unable to flee. Slowly, neighboring tentacles bend around the prey and hold it firmly in place.

Glands on the tips of the tentacles then secrete a digestive enzyme, and within a few hours the insect is transformed into an unrecognizable mass. Unlike most other carnivores, sundews, with their extraordinarily powerful digestive system, often ingest the prey in a matter of hours.

Most American species are small, jewel-like plants, rarely more than four inches high. But the giant *Drosera regiae* of South Africa has leaves two feet long, and is capable of devouring small animals in its trap.

In Australia there are several species of pygmy sundews, miniature plants with an unusual method of reproduction. While other sundews blossom and produce seed like other flowering plants, pygmy sundews instead produce spore-like gemmae. These gemmae are formed in the center of the plant and resemble eggs in style and purpose. In fact, at the end of a growing season, the whole plant looks like a bird's nest full of eggs.

The sundew, found on every continent, usually grows in peat bogs and acid soil. A particularly hardy plant, it is often the first new growth to appear on a fire-ravaged forest floor. During the winter it recedes to a small bud above the ground, renewing its

Close-up of *Drosera* with its tentacles holding captured prey.

growth in the spring. The sphagnum moss it grows in, however, does grow during the winter, so the plant often grows in layers of sundew/moss/sundew/moss, tied together by long stems that root between generations.

The hardiness of the sundew makes it ideal for the home grower. It is also particularly easy to reproduce. Flowering and seeding readily, the plant is self-fertilizing. Blossoms usually open for several hours in the morning, with each flower blooming only for one day.

Sundews also reproduce asexually, through leaf blade reproduction. The base of each tentacle contains cells capable of producing new plants. To reproduce by this method, remove a leaf blade from the plant, using a very sharp knife. Success is variable, depending upon the age of the leaf, so select a young and vigorous one. Now slice the leaf into as many pieces as possible. Scatter the pieces on sphagnum moss, dampened several days in advance, and keep it moist constantly. Invert a plastic bag over the pot of sphagnum, perhaps propping it up with a stick to form a tiny greenhouse. Keep it in the sun, and within a month you'll see a tiny new sundew growing from each piece of leaf. After the plants are larger, transplant them if you want and grow them normally.

Seedling *Drosera* with seedling *Dionaea muscipula. Dionaea* has just captured a gnat. Note size in relation to the dime at bottom of photograph.

Drosera adelae

Location: Australia, particularly North Queensland near Rockingham Bay, and the Hinchbrook Islands.

Description: Distinctive, lance-shaped leaves, seven inches long and very soft. May look like *Pinguicula* to the uninitiated.

Cultivation: Easy to grow, but difficult to obtain. Will not withstand frost.

Reproduction: Seeds, leaf cuttings, and cuttings from adventitious roots near the soil surface.

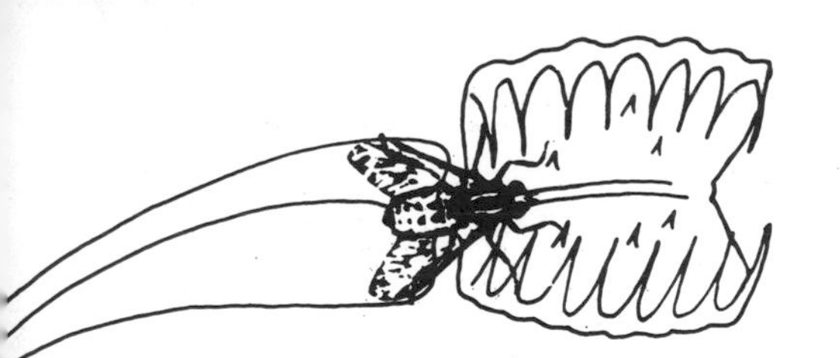

Drosera anglica

Location: Western North America and eastern Canada.

Description: Three inches in diameter with short, paddle-shaped leaves. Leaves often upright. The photo (opposite) is of a group of young plants which have not quite taken final form. Leaf is normally slightly more elongated.

Cultivation: Easy to grow, but not readily available. Will withstand frost.

Reproduction: Seeds and leaf cuttings.

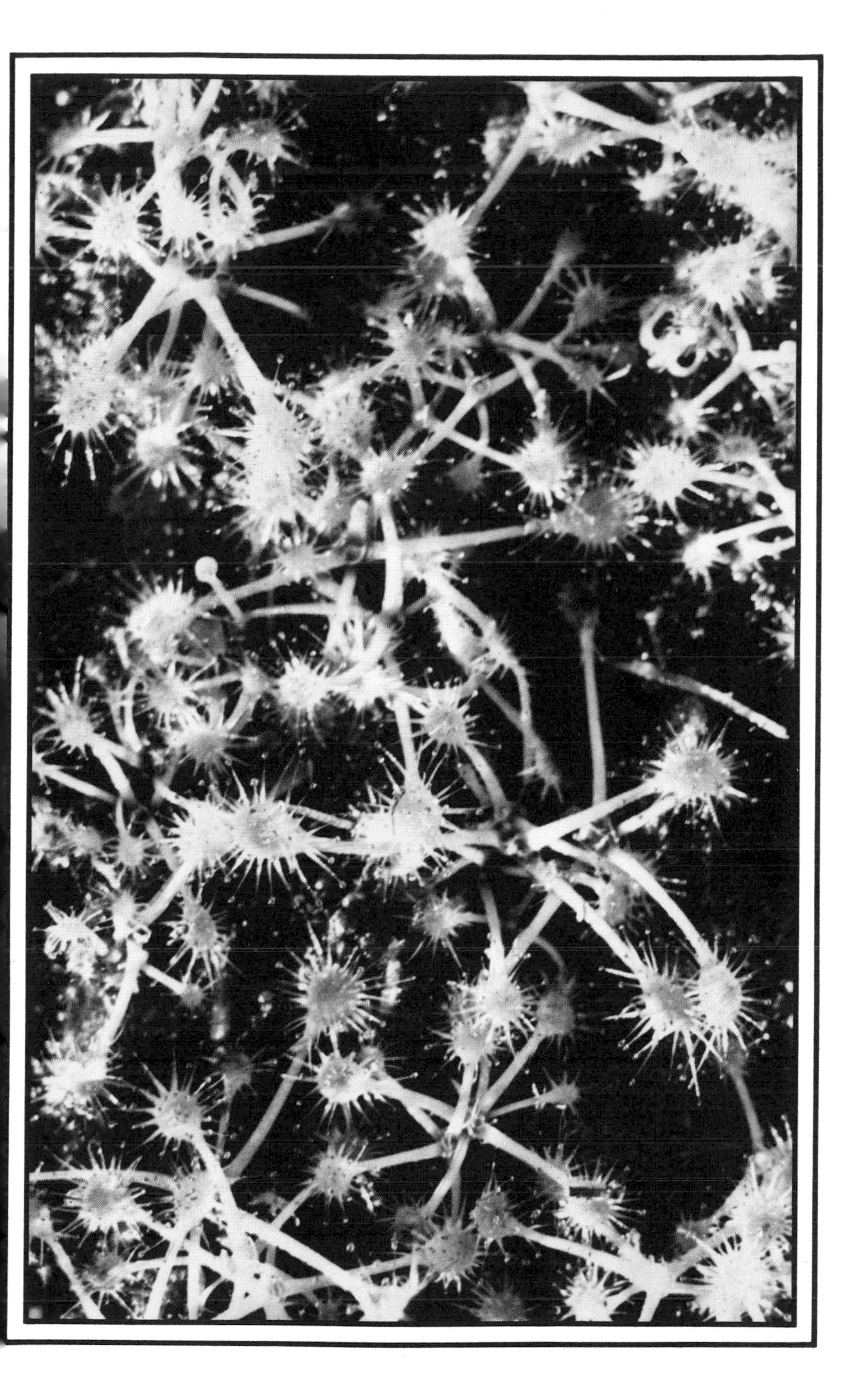

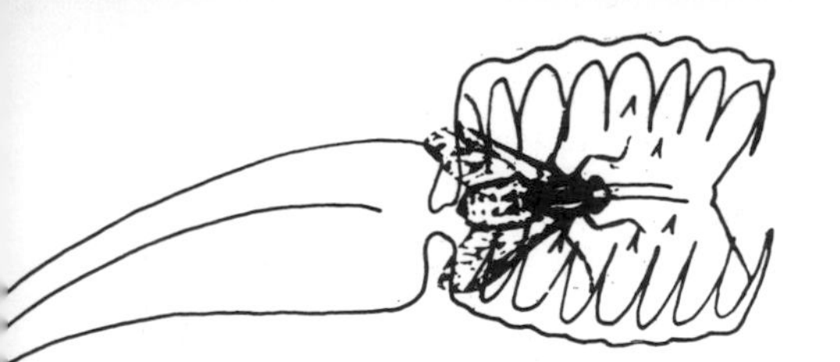

Drosera binata

Location: Australia, New South Wales, and New Zealand.

Description: May be short and stout, with red-tipped tentacles, as in the photograph, or, if grown in shade, has green and long, even drooping leaves. Flowers are white.

Cultivation: Easy to grow; available commercially, will not withstand frost.

Reproduction: Seeds, root and leaf cuttings. Very prolific, practically a weed.

Drosera binata shows three forms. The basic type is simply one stem split into two blades, like a T. Another form, variation *dichotoma*, splits again, as in the photograph, producing multiple forks. The third form goes wild and splits again and again, looking like a maze of leaves. This is called *multifida*. The plant may not show its final form until after the first year, so watch for surprises.

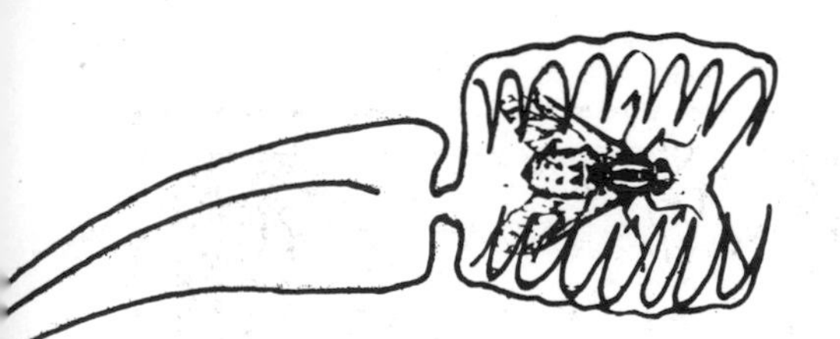

Drosera capensis

Location: South Africa

Description: Four inches in diameter, with long, paddle-shaped leaves. Produces a long stem sheathed with dead leaves and tipped by the active plant. Bushy and trailing, putting down roots where it touches the ground.

Cultivation: Easy to grow; available commercially. Will not withstand frost.

Reproduction: Seeds, root and leaf cuttings; also, by cuttings from the growing stem.

This species is a particularly active sundew. Often the leaf blades, as well as the tentacles, curl around a victim to insure its capture.

Drosera capillaris

Location: North America

Description: Two inches in diameter, with reddish-purple flowers on a tall stem.

Cultivation: Easy to grow; available commercially. Will withstand frost.

Reproduction: Seeds and leaf cuttings.

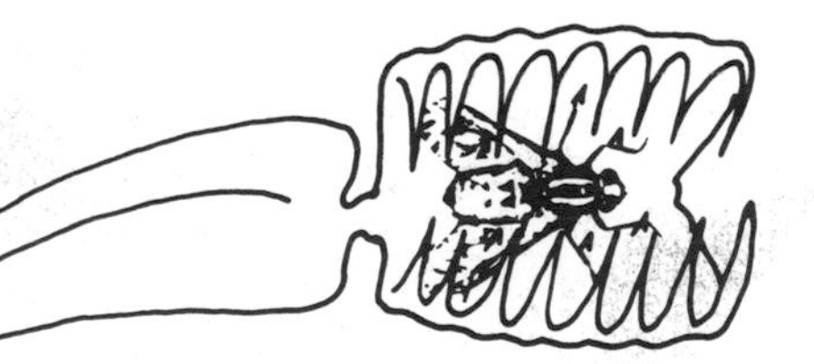

Drosera filiformis

Location: North Atlantic seacoast, from New Jersey to Florida, and along the Gulf Coast to Mississippi.

Description: About two feet tall, looks like dew-covered grass blowing in the wind. Plant in photograph is still young, only inches high. Species has two varieties: one, always green; the other, more pigmented, with red-tipped tentacles and dark purple flowers.

Cultivation: Easy to grow, available commercially. Will withstand frost.

Reproduction: Seeds and leaf cuttings.

Drosera intermedia

Location: North America, especially the boglands of California, New Jersey, and North Carolina.

Description: Up to four inches high, leaves jutting up in a circle, forming a cup at the base. Produces winter buds during the dormant season.

Cultivation: Easy to grow, but not readily available. Will withstand frost.

Reproduction: Seeds and leaf cuttings.

Drosera peltata

Location: Australia, particularly in the swamps and heaths behind Mount Kina, New South Wales.

Description: Six or seven inches high, growing erect in a sort of vine. Numerous shield-shaped (peltate) leaves. Flowers are white.

Cultivation: Difficult to grow or obtain. Will not withstand frost.

Reproduction: Seeds and tubers.

Peltata belongs to a group of sundews which have tuberous roots producing underground tubers. These tuberous *Droserae* form "dropper" roots from the central tuber. These droppers then form new tubers at their tips which store food over the dormant season and produce new plants.

The life cycle of the peltate species, typical of tuberous *Drosera,* is quite unusual. Its dormant period is during the summer, the dry season. The plant begins to grow from its underground tuber in late autumn and soon forms a rosette of leaves. Then, gaining strength, it produces its distinctive tall stem, topped in the spring with white flowers. After seeding in early summer, the plant becomes dormant, with the portion above ground gradually dying.

The Australian origin of the plant, together with its uncommon life cycle, often produces cultivation problems. When the plant is moved to the northern hemisphere, where the seasons are reversed, it experiences a kind of jet lag. In reacting, the *peltata* may fail to produce the dropper roots and new tubers, thus being unable to store food or reproduce. In that case, the plant will eventually die.

If the plant successfully establishes itself, it will grow. Its dormant summer requires rather cool temperatures. Remember that summer is also its dry season. Stop watering the plant when the stem dies, and don't begin again until you see new activity in autumn.

Drosera rotundifolia

Location: North America, Europe, and Asia. This is the predominant North American species, found especially on the seacoast belt from New Jersey down to Florida, and along the Gulf Coast to Mississippi.

Description: An inch in diameter, flat with round leaf blades, bright red growing in the sun. Flowers, open in the morning, are white or pink, sometimes yellow, mauve, or purple. Forms winter buds to survive freezing.

Cultivation: Easy to grow; available commercially. Withstands frost.

Reproduction: Seeds and leaf cuttings.

Medicinal use: Potion used to cure coughs, whooping coughs, and respiratory diseases.

"There is a usual drink made thereof with aqua vitae and spices frequently and without any offence or danger but to good purposes used in qualms and passions of the heart."

– Nicholas Culpeper,
The Complete Herbal
and English Physician, 1642

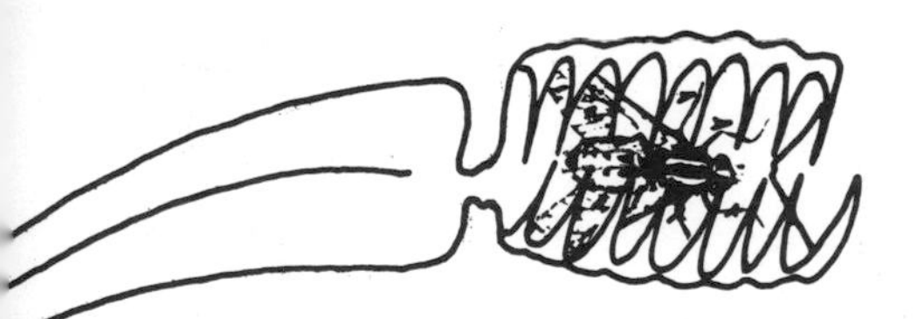

Drosera schizandra

Location: Australia, only in North Queensland in the region of Mount Bartle Frese.

Description: Four inches in diameter, broad leaves with a center notch, except for plants grown in cultivation, which may not notch.

Cultivation: Easy to grow, but not readily available. Will not withstand frost.

Reproduction: Seeds and leaf cuttings.

Drosera spathulata

Location: Asia and Australia.

Description: Two inches in diameter, with glands bright red when growing in strong sun. Tentacles grow over the entire leaf surface, which tapers into the base. This produces the spatulate shape which gives the plant its name.

Cultivation: Easy to grow; available commercially. Will not withstand frost.

Reproduction: Seeds and leaf cuttings.

Byblis

Picture a giant sundew, growing about two feet tall. It looks like a leafless shrub with a strong woody base. Its limbs are covered with drops of sticky fluid, deadly to its prey. Actually, it's not *Drosera*, but its distant relative, *Byblis.*

Growing in western Australia, *Byblis* may have given rise to those rumors about man-eating plants. It is often found intertwined in clumps, forming a strong, sticky hedge. As insects are caught on the gluey surface, small animals such as frogs and lizards come after the insects. *Byblis* captures and digests them, too. While plants grown in a greenhouse don't attain full size, the larger, natural Australian plants are supposedly capable of capturing rabbits or squirrels. This is one carnivore for which the name "insectivorous" is clearly inadequate!

Byblis gigantea and its only actual relative, *Byblis linifolium*, grow in drained soil rather than the boggy conditions most other carnivores prefer. My *Byblii* are thriving in a mixture of equal parts sphagnum moss, pearlite, and sand, kept moist. The plant needs less moisture because it has a thick root for storing water.

The thickest part of the central root can also be used for propagation. When repotting, take pieces about one inch long and place them on top of some damp sphagnum. Cover the pot with plastic to help keep it moist. In about a month you should have a new plant.

Constantly in blossom, *Byblis* produces flowers that are only open one day before they die. Pollination is difficult, requiring vibration to release the pollen. When seed is produced, it takes from three months to two years to germinate. Better stick to the root cutting method!

Probably the oddest thing about *Byblis* is a small insect that lives on the plant, apparently immune to its dangers. Not only does this strange insect live on the carnivore undisturbed, it actually feeds on the juices of the plant's freshly captured prey! When grown in America, *Byblis* is also susceptible to aphids, a pest that has not been known to bother the plant in Australia.

Drosophyllum

Having *Drosophyllum* in the house is something like growing your own flypaper. Charles Darwin wrote, "the . . . fact is well known to the villagers, who call the plant the 'fly catcher' and hang it up in their cottages for this purpose."

Drosophyllum has no moving parts. It depends totally upon the droplets of liquid on each tentacle to bring down its prey. The secretion is different from that common to *Drosera*, however. The sticky stuff doesn't draw out, chewing gum style, between the tentacle and the victim, holding the victim captive. Instead, the liquid is released from the plant when an insect brushes against a tentacle. The insect, carrying a drop of the liquid that has stuck to it, moves away from the first tentacle and on to the next, where the scene is replayed, until the insect is literally swamped in the stuff and drowns under the weight. The dead prey comes to rest against a layer of sessile glands covering the surface of the leaf. These glands secrete digestive enzymes, which reduce the victim to its basic organic components.

Drosophyllum is a pretty plant. The glands are red, giving the entire plant a warm, rosy hue. The flowers are a bright sulfur yellow, and grow quite prolifically in the second season.

Quite unusual for a carnivore, the plant grows in dry, rocky soil, while the air is humid. It can be found near Morocco, and adapts quite well as a houseplant. It is unavailable commercially but can, with difficulty, be secured from botanical gardens in Morocco in seed form.

Drosophyllum attains a height of up to two feet tall, growing in the form of a single stalk, never in clumps. Apparently, the individual plants are mutually inhibitive and cannot be grown together. If you're growing them from seed, separate your plants when they're several inches high.

Drosophyllum should be kept in well-drained soil mixed with

Close-up of *Drosophyllum* with trapped prey. Note sessile glands.

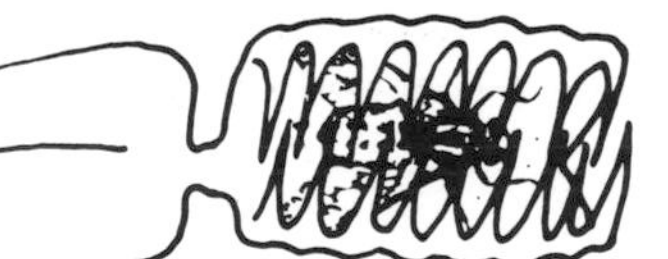

pearlite or sand. Remember, it's not a bog plant at all. Apparently a biennial, *Drosophyllum* blossoms during its second season, sets seeds, and then dies. Be sure to hold on to the seed of each successive generation.

Before planting, take the mature seed and scrape it lightly on sandpaper to scratch the seed coat. Sow them in equal parts of sphagnum moss and pearlite, keeping them moist but not wet. While seedlings, the plants are susceptible to damping-off, so be careful.

Drosophyllum lusitanicum.

Butterwort – *Pinguicula*

There's a lot of folklore attached to the butterwort plant, but not the gory tales you might expect. In Scandinavia, the leaves are used as a medicine, rubbed on the open sores of animals to make them heal faster. This practice is not based simply on superstition; there is actually an antibiotic in the plant's digestive fluid that keeps captured insects from decaying.

The plant is also commonly used as a milk coagulant in home dairy recipies. There's no known scientific basis for this effect, however; it just works.

Butterwort gets its Latin name, *Pinguicula*, from *pinguis*, meaning fat, referring to the greasy feel of the leaves.

The entire surface of each leaf is covered with two types of glands. One secretes an oily mucilage, which increases and drowns the prey when an insect is caught in it. Then, the other releases an acid liquid that digests the victim, which is finally absorbed into the leaf. The digestion process is triggered by any captured object containing nitrogen; thus, the plant feeds not only on insects but on seeds and pieces of leaf as well.

The only creatures vulnerable to the butterwort are tiny insects such as gnats; the plant can't capture anything larger. Victims are attracted to the butterwort by a fungus-like odor. When the prey is caught in the liquid secretion, the leaves move slightly, rolling upward to form a shallow cup for the digestive fluid.

Most species of *Pinguicula* are similar in appearance. The plant consists of a rosette of decumbent leaves, very soft and easily bruised. Their bright-green color becomes yellowish in strong sunlight. Depending on the species, butterwort flowers may be blue, purple, or yellow, growing singly on a long stalk. Flowers on the tropical species are grand and showy enough to be cultivated in greenhouses just for that purpose.

Butterworts will thrive in the same conditions as the other

A *Pinguicula* blossom.

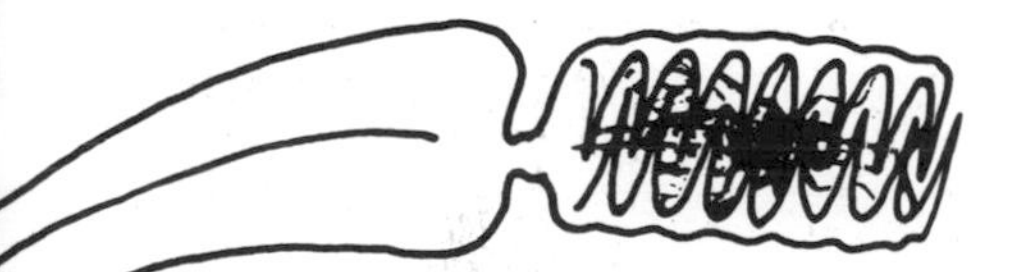

carnivores. Come winter, some species may regress into a tight bud. If the plant produces two or more buds, separate them and each will become a complete plant in the spring.

Plants can also be reproduced by seed, or by leaf propagation. For the leaf method, premoisten a pot of sphagnum moss. Make sure the moss is only damp and not wet, since there can be danger of rot in the first few weeks. Gently peel some leaves off the butterwort, being careful to remove the whole leaf, including the white base. Now lay the leaf on the sphagnum, tucking the base into the moss, and cover the pot with plastic. Seal it up and put it in good light. When the new plants have several leaves, you can treat them as mature plants.

Pinguicula caudata

Location: Mexico and Central America

Description: Up to five inches in diameter, with leaves yellow-green to reddish, from shade to sun. Flowers are generally purple, and very showy. The plant is capable of taking many forms, changing leaf shape and flower color from plant to plant, and on the same plant from season to season. Several butterworts have been discovered and named, and later found to be *Pinguicula caudata* in disguise.

Cultivation: Easy to grow; available commercially. Will not withstand frost.

Reproduction: Seeds and leaf cuttings; also, offshoots.

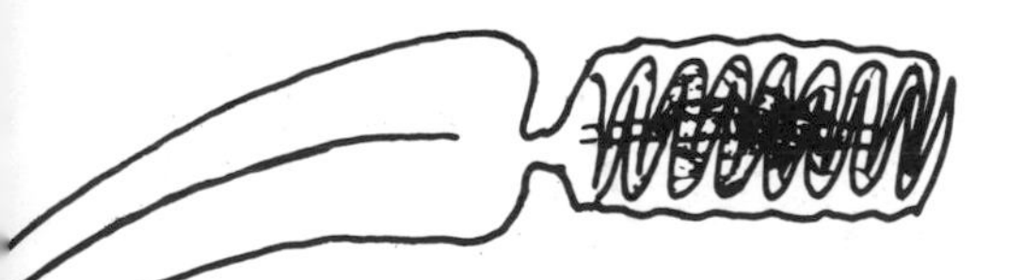

Pinguicula mexicana

Location: Mexico.

Description: Up to four inches in diameter. Flowers deep purple.

Cultivation: Easy to grow; available commercially. Will not withstand frost.

Reproduction: Seeds, leaf cuttings, and offshoots.

NOTE: This plant was found and popularized by J. N. Giridlian, a nurseryman in California. It may be only a form of *Pinguicula caudata.* I feel, however, that until finally determined it should be treated as a species.

Pinguicula planifolia

Location: West Florida, west to Mississippi.

Description: Three inches in diameter; leaves reddish-purple if grown in good light. Flowers are purple.

Cultivation: Easy to grow, but not available commercially. Will withstand light frost.

Reproduction: Seed, leaf cuttings, and offshoots.

Pinguicula primuliflora

Location: Florida, Georgia, and Mississippi.

Description: Three or four inches in diameter. Blue, white, or yellow flowers.

Cultivation: Easy to grow, but not available commercially. Will withstand frost.

Reproduction: Seed, leaf cuttings, and offshoots.

3. *Passive Traps*

They lie in wait – jewels of nectar just out of reach beyond slick tunnels to death. Their victims try to reach the nectar treasure but fall prey to the ingenious traps.

The traps don't move; they don't need to. And since they're never triggered, they're never unprepared for their next victim.

All passive plants are variations of the pitcher plant. While sweet nectar entices the insect, a one-way tunnel leads it to its death. Traps range from delicate little cups to pitchers holding several quarts of fluid, ready to digest giant tropical rats.

Pitcher plant – *Sarracenia*

Pitcher plants use neither movement nor adhesive to catch their prey. Instead, they rely totally on the ingenious design of their traps, which display incredible variety as well as some fascinating structural features.

The basic trap resembles a balloon, just slightly inflated, with a long wing attached. Like a pitcher, the trap is a vessel capable of holding a liquid – in this case, digestive fluid. Before the victim gets there, however, it goes through what amounts to an obstacle course to insure its capture and reduce the possibility of escape.

Let's walk with the prey for a while. The insect is attracted by sweet nectar secretions on the outside of the trap, concentrated on the wing, and leading to the mouth. At the mouth of the trap is a flap covered with hairs pointing downward, into the pitcher. It's a very colorful and attractive area, and fairly easy going moving down the hairs. But it's one way only, since, after the insect passes through, the hairs are spikes at the insect's back.

Once past the hairs, the insect can only try going down or around since it can't go back. It could try getting out over the lip around the other side, but there's a precarious overhang that's slick as glass.

The only way to go now is down. Suddenly the whole area becomes very slippery, and footing is difficult. It's not at all easy to climb, and the victim is forced into a region of glass-like hairs. Here, it's very hard to remain stationary, but it's impossible to go up, and there's a pool of water below.

The prey is doomed, for the liquid is not water. It contains a wetting agent which reduces surface tension, making it impossible for the insect to crawl along the surface. The victim quickly drowns. Digestive enzymes in the liquid disintegrate the soft matter of the insect, which is absorbed by the plant.

While the exact details vary somewhat among the different

Sarracenia purpurea with freshly captured prey.

species of pitcher plants, the basic scenario changes very little. The mound of skeletal remains found in the bottom of the pitcher is mute testimony to the trap's success.

On each plant there is a rosette of pitchers and a single, long-stemmed flower. Each pitcher has a protective cap growing over it to keep rainwater from flooding the trap. The shape of the pitcher varies greatly from species to species, and even within the same species. But the flowers remain identical, varying only in color. They blossom above the plant, turning downward to hang like a bell.

Like most other carnivores, pitcher plants are happy in acid soil or moss and high humidity. Chronically low humidity may cause the edges of the pitcher cap to brown. In most species, the size of the outer wing in relation to the pitcher is a good indicator of proper light. A wing smaller in profile than its pitcher means the plant is getting enough light. Some varieties also take on a colored vein pattern in good light. *Sarraceniae* are especially sensitive in their dormant season. Don't force them to grow; they may die.

During the growing season, the plants leave a horizontal, underground runner called a rhizome. When this rhizome is several inches long, you can cut a two-inch piece and plant it to form a new plant. Often rhizomes are sold in bulb form much like tulips.

Sarraceniae also reproduce easily from seed and often form natural hybrids in the field. If you find two different species growing in the same vicinity, there may also be some plants nearby that combine the features of both. You can do it yourself, simply by taking pollen from one species and applying it to the pistil of the flower of another species.

All but one species of *Sarraceniae* are native to the southeast. In fact, they can all be seen at one particular location, some in incredible numbers. The station is by U.S. Route 90 at the Mississippi-Alabama state line.

Sarracenia flower.

Sarracenia drummondii

Location: Coastal Alabama and Florida panhandle.

Description: Three feet tall; strikingly beautiful. White top, with green and red veins. Flowers are red.

Cultivation: Difficult to grow; not readily available. Will withstand light frost.

Reproduction: Seeds or rhizome division.

NOTE: *Sarracenia drummondii* has recently been changed to *Sarracenia leucophyla.*

Sarracenia flava

Location: Coastal Alabama, Georgia, and the Carolinas.

Description: Three feet tall; yellow-green with yellow flower. Sometimes red-veined.

Cultivation: Easy to grow; available commercially. Will withstand light frost.

Reproduction: Seeds and rhizomes.

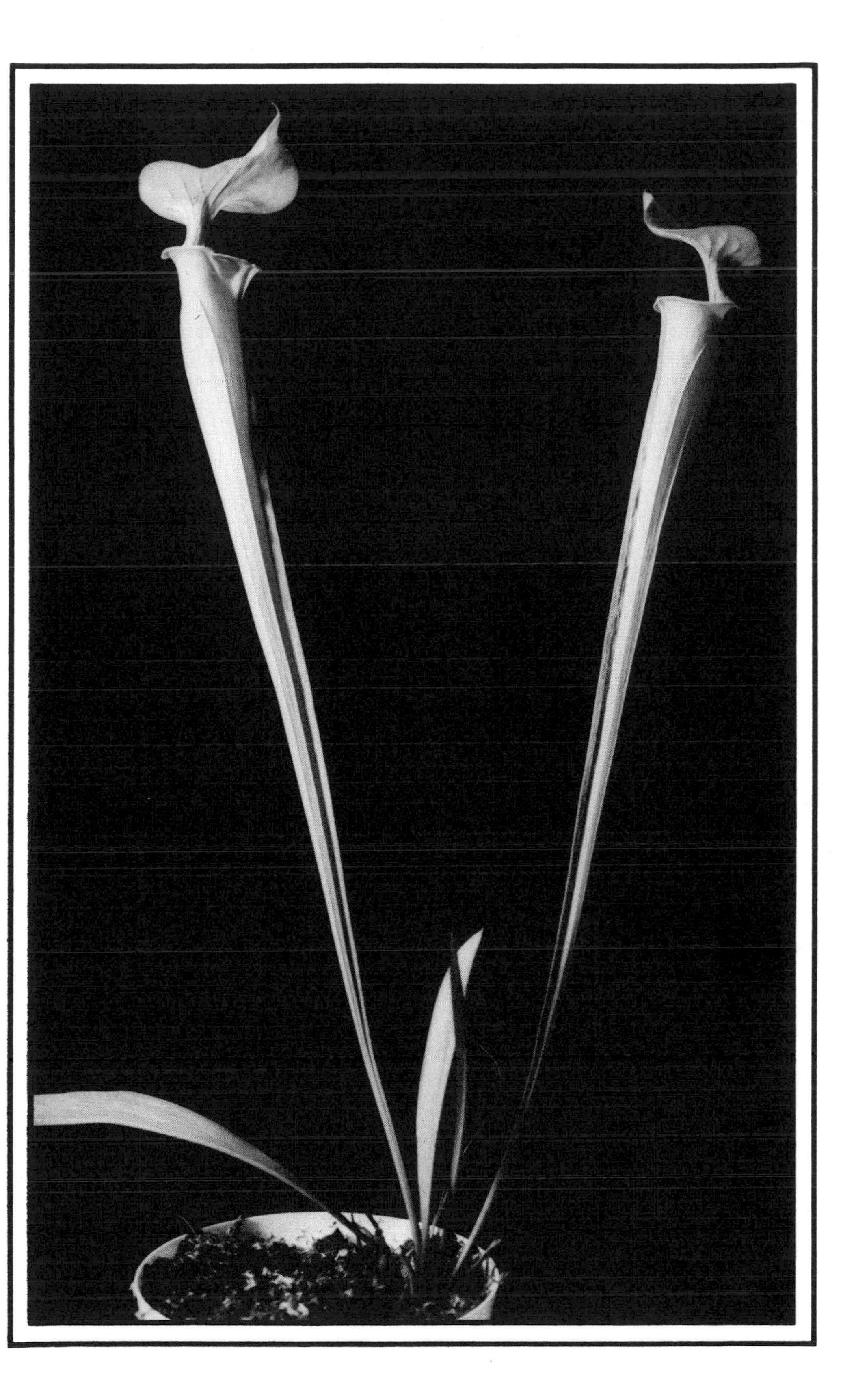

Sarracenia psittacina

Location: Southern Mississippi, Alabama, and Georgia.

Description: Recumbent plant seven inches in diameter, turns red growing in sun. Hood has clear spots. Flower is red.

Cultivation: Easy to grow, commercially available. Will withstand light frost.

Reproduction: Seeds and rhizomes.

Sarracenia purpurea

This plant has two subspecies: *Sarracenia purpurea venosa*, shown in the photograph; and *Sarracenia purpurea gibbosa.*

Sarracenia purpurea venosa

Location: Coastal Mississippi, Alabama, North Carolina, and New Jersey.

Description: Recumbent plant, about eight inches in diameter. Flowers are red.

Cultivation: Easy to grow; commercially available. Will withstand light frost.

Reproduction: Seeds and rhizomes.

Medicinal Use: A hormone found in its roots is now being studied as a possible cancer cure.

Sarracenia purpurea gibbosa

Location: Maryland, Delaware, and New Jersey, north to subarctic Canada and west past the Great Lakes.

Description: Slightly upright, about eight inches in diameter. Pitchers are longer and more narrow, with a larger flap. Flowers are red.

Cultivation: Somewhat difficult to grow; commercially available. Will withstand frost.

Reproduction: Seeds and rhizomes.

Medicinal Use: Potion of roots and leaves used as abdominal cure-all for stomach, diuretic, and menstrual complaints. At one time, said to ward off smallpox.

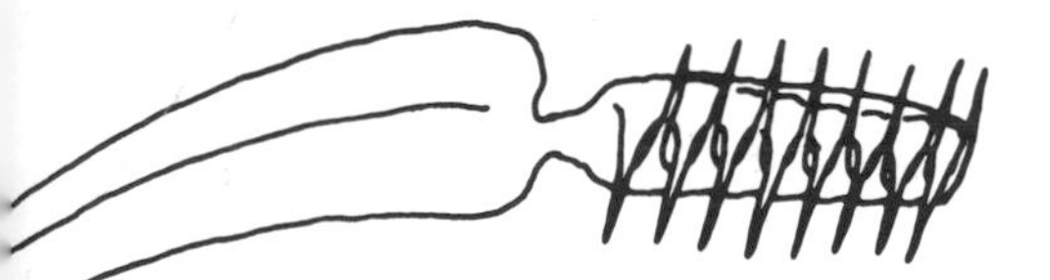

Cephalotus follicularis

Cephalotus is a small fuzzy jewel with a gory-looking mouth. The pitchers are an odd combination of plush pubescence and deadly appearance.

The plant is attractive not only to people, but to insects, particularly ants. The traps, quite efficient, have downward-pointing hairs from the top of the lid to about halfway down the pitcher, with scattered nectar glands. Victims are digested by plant enzymes as well as bacteria.

Unlike most carnivores, *Cephalotus* grows year round and produces different leaves in different seasons. During summer and autumn, the plant grows simple flat leaves, small and oval-shaped. During winter, however, it grows the traps, which will be mature and ready for the first insects of spring.

Cephalotus grows in dense clusters, then produces offshoots that grow new plants. In the field, it is found covering the ground with a lush layer of pitchers and plants.

Green with a red tinge, the mouth of the pitcher is lined with red shiny ridges that look like a necklace of animal teeth. The pitchers are quite small, rarely more than two inches long. The other leaves are green on top, red on the bottom. Clusters of white flowers extend up a tall stalk in the spring.

While many people grow *Cephalotus*, real success is rare. Occasionally the plants flower and flourish in captivity, but vigor seems sporadic. There's some talk of a need for salt in their diet, but I can't seem to pin down the rumors, so I don't recommend such action. They do grow better in peat than in sphagnum. Cover the peat with a layer of sphagnum, though, to keep down the splash if you water from the top. Once established, *Cephalotus* will withstand frost.

Heliamphora

Many carnivorous plants grow in limited geographical areas, but none so remote as Mount Drivoa, Venezuela. There grow the pitcher plants known as *Heliamphorae*, perhaps the rarest of the carnivores.

When found, the plants look like graceful green funnels growing out of the moss. Long red stems hold delicate white blossoms high above the traps. The pitcher is a leaf, rolled up and seamed at the ends. Over the top grows an abbreviated cap, a small, spoon-shaped nipple.

Most pitcher plants provide themselves with some sort of protection from rainwater, which would otherwise flood them till they overflowed and collapsed from the weight. They have a hood for this purpose, but the tiny rain cap of *Heliamphora* is ineffective. The plant has a clever feature to solve the problem. A slit in the funnel acts as a drain to regulate the water level, so that the plant can thrive in the almost constant rain of its natural habitat.

Attracting insects with a sweet nectar, *Heliamphora* employs alternate regions of hairy and slick surfaces within the pitcher to ensure capture. The plants do not have digestive enzymes, but depend instead on bacterial action to decompose their prey. The plant then absorbs its food.

Heliamphorae are very difficult to obtain, and are so rarely grown in greenhouses that it was thought they could not be cultivated. As their availability has begun to increase, this myth has been dispelled. The plant is grown in the standard wet sphagnum moss and high humidity. It does, however, adapt very slowly to captivity, and may react with lack of blossom, lack of pollen, or partially developed juvenile pitchers before it adapts successfully.

Heliamphora minor.

Heliamphora minor

Location: Mount Drivoa, Venezuela.

Description: A short funnel with a seam. White, showy flowers on a tall stem. About four inches tall.

Cultivation: Easy to grow, but very difficult to obtain.

Reproduction: Seeds and rhizomes. May not reproduce sexually in captivity.

Heliamphora heterodextra

Location: Mount Drivoa, Venezuela.

Description: A tall thin funnel topped with a tiny spoon-shaped cap.

Cultivation: Easy to grow, but very difficult to obtain.

Reproduction: Seeds and rhizomes. May not reproduce sexually in captivity.

Heliamphora heterodextra.

Darlingtonia californica

Darlingtonia, the cobra plant, looms like a hooded snake, head held high and about to strike. It even has "fangs," which unwary insects follow right into the mouth of the trap.

Darlingtonia traps its victims in the same way as other pitcher plants. Nectar glands cover the outside of the tube, attracting insects. Flying insects have an easy landing pad on the fishtail-shaped fangs, which are like a tongue leading to the mouth of the trap. Insects crawling up find this ramp covered with hairs, pointing upward for easy access.

Once the victim is over the edge of the lip and into the mouth, the hairs point downward, encouraging the insect to proceed further inside. Escape is difficult, since the hairs form a barrier at the insect's back once it is inside. To further confuse things, the head of the cobra is spotted with false openings, colorless cells like windows that mislead the victim, preventing it from finding its way out.

The prey ends up in the base of the pitcher, drowned in the liquid in the bottom. The victim deteriorates by bacterial action, since there are no digestive enzymes in the fluid. Sometimes there's an unpleasant odor around plants that have a lot of decaying insects inside.

Darlingtonia is closely related to *Sarracenia* and *Heliamphora,* Apparently, when the plants were evolving, *Darlingtonia* was cut off from the eastern-based *Sarracenia* by the upheaval of the Rocky Mountains. The cobra plant developed into its present form in a narrow area in northern California and Oregon, hence its full name, *Darlingtonia californica*. You may find it occasionally referred to as *Chrisamphora californica*, but *Darlingtonia* is its official name.

Strikingly beautiful, tall, and graceful, *Darlingtonia* grows about three feet high in its natural habitat. Homegrown plants are usually smaller. In the spring, a single red, drooping flower blooms high above the plant. The sun reddens the tops of the plants during the season, so that by the end of summer, the heads are a dark red color.

The plant thrives in very wet conditions; sometimes it can be found almost floating in streams and bogs. Used to the cool climate of California, *Darlingtonia* dislikes heat, especially warm nights. This and the extreme wetness may be problems if you try growing cobras at home.

If you successfully raise a *Darlingtonia*, you'll be rewarded by seeing the plant's exceptional growth pattern. The development of each individual plant dramatizes the evolution of the species. Very young plants often grow primitive forms of pitchers. Some look like *Heliamphora* with the cap extended into a long tail. Others look like a rolled leaf. As the plant matures, new pitchers have the long tail forming more and more into the cobra's distinctive head. In rare cases, this trait can also be found in other types of pitcher plants, but *Darlingtonia* is dependable for it.

Like the *Sarracenia*, new cobra traps make their first appearance facing the center of the plant. The young traps are tall and flat. As each new trap grows, however, it twists around until the head faces outward, and puffs up, giving an almost threatening appearance.

Magnification of *Darlingtonia californica* hood. Note transparent sections.

Nepenthes

According to folklore, the exotic-looking *Nepenthes* can cause rain and cure bed-wetting! These magical powers are attributed to the liquid that fills the pitcher-type trap. Referred to in Homer's Odyssey, it was believed to induce forgetfulness. Today, the sterile, antibiotic fluid of the as-yet-unopened immature pitcher is used as a natural medicine and thirst quencher by people from Indonesia to Madagascar.

Nepenthes often grows in the jungle as a vine, clambering through trees or running along the ground. The full grown plants are several feet long, with large, simple, flat leaves tipped with a tendril. The pitchers extend from several of the tendrils, which support and attach each pitcher to a leaf. The pitchers, usually with a capacity of about a half pint, sometimes grow large enough to hold several quarts of liquid.

When insects lean over the lip of the trap to get at the nectar inside, they fall into the pool of liquid, slipping on the slick edge of the lip. A walking insect, which might be able to climb out, has its foot pads coated by a loose, dusty wax and invariably falls back into the liquid. A flying insect, wings too wet to fly, is also trapped. Once inside the pitcher, the prey ultimately drowns in the fluid and is digested by plant-produced enzymes.

Popular with Victorian horticulturists and botanists, *Nepenthes* has been hybridized, adding new pitcher shapes to the remarkable range that occur in nature. There are some twenty species, as well as unknown quantities of hybrids, each having a male and female plant, distinguishable only by their blossoms.

A fascinating aspect of the plant is the development of the pitcher. At first there is only a swelling on the tip of the tendril. Slowly it takes on the shape of a flattened pitcher trap with its hat smashed down flat. Suddenly the tip, no matter how many curls may be in the tendril, becomes upright. Then the pitcher grows larger. Finally it begins to puff out from its flat shape into the round pitcher shape. The pitcher is now about one-third full of its liquid and very heavy, causing the leaf to droop if its tendril is not wrapped about a supporting tree branch. In the formation of the

final shape, the cap begins to rise off, revealing the cottony, fuzzy plug that has kept the liquid sterile. As the cap rises into place the lip begins to roll and form its distinctive teeth, resulting in a fully functional trap.

Nepenthes is happiest in clayish soil or sphagnum. I use sphagnum because it holds water so well, and these plants need a lot of water. While it's easy to grow, the plant requires very high humidity to produce pitchers. Most growers use enclosures or misting to approximate the steamy jungle conditions of the plant's native habitat. I also pour a mild solution of fertilizer over the leaves occasionally during the growing season.

Nepenthes plants are either male or female; the male plants blossom with only the pollen, the female with only the pistil. But since the plants rarely blossom with the right sex on the right plant at the right time, collectors don't multiply *Nepenthes* by seed. The most popular method of reproduction is stem cuttings. Since more pitchers are produced by young plants, use cuttings from your mature plants to start new growths. Snip off sections of the main stem, one leaf per section. Cut off half the leaf. Dip both ends of the stem in Rootone, and insert it into a pot of moist sphagnum moss. The rooting piece is susceptible to damping-off, fungus, and a host of other dangers, so make sure the moss isn't too wet. Now cover the pot with a plastic bag to cut down on transpiration. When the new plant has two or three leaves, the bag may be removed.

Immature *Nepenthes* trap beginning to show characteristic shape and cap.

Nepenthes ampullaria

Location: Malaya, Sumatra, Borneo, New Guinea.

Description: Tall vines, or short, in clusters, Pitchers three or four inches high and green. Often forms pitchers directly from roots or underground stems.

Cultivation: Typical *Nepenthes* culture; not commercially available. Will not withstand frost.

Reproduction: Stem cuttings, or seed with difficulty.

Nepenthes dicksoniana (commercial hybrid)

Description: A hybrid created by Vietch Nurseries, France. Tall vines. Pitchers six or eight inches high, with wide flaring mouths. Green, mottled with red.

Cultivation: Typical *Nepenthes* culture; available commercially. Will not withstand frost.

Reproduction: Stem cuttings, or seed with difficulty.

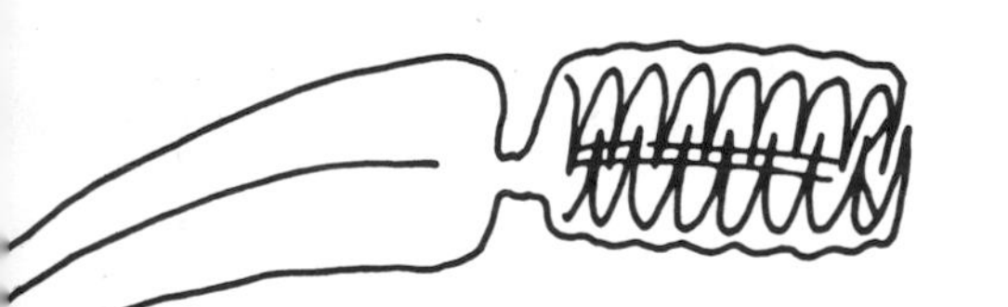

Nepenthes gracilis

Location: Borneo, Malaysia, and Sumatra

Description: Smaller than other *Nepenthes*, leaves narrow and about four inches long. Forms clumps before growing tall. Pitchers are tall and thin.

Cultivation: Typical *Nepenthes* culture, but will withstand cultivation errors. Very prolific pitcher formation, especially when young. Will not withstand frost.

Reproduction: Stem cuttings, or seed with difficulty.

Nepenthes hookeriana (natural hybrid)

Location: Borneo

Description: Natural hybrid between *Nepenthes ampullaria* and *Nepenthes rafflesiana.* Tall vine with short, fat pitcher about four inches high. Green mottled with red. Pitcher in photo (opposite), having just opened, is slightly immature; its hard lip has not fully formed.

Cultivation: Typical *Nepenthes* culture; available commercially. Will not withstand frost.

Reproduction: Stem cuttings, or seed with difficulty.

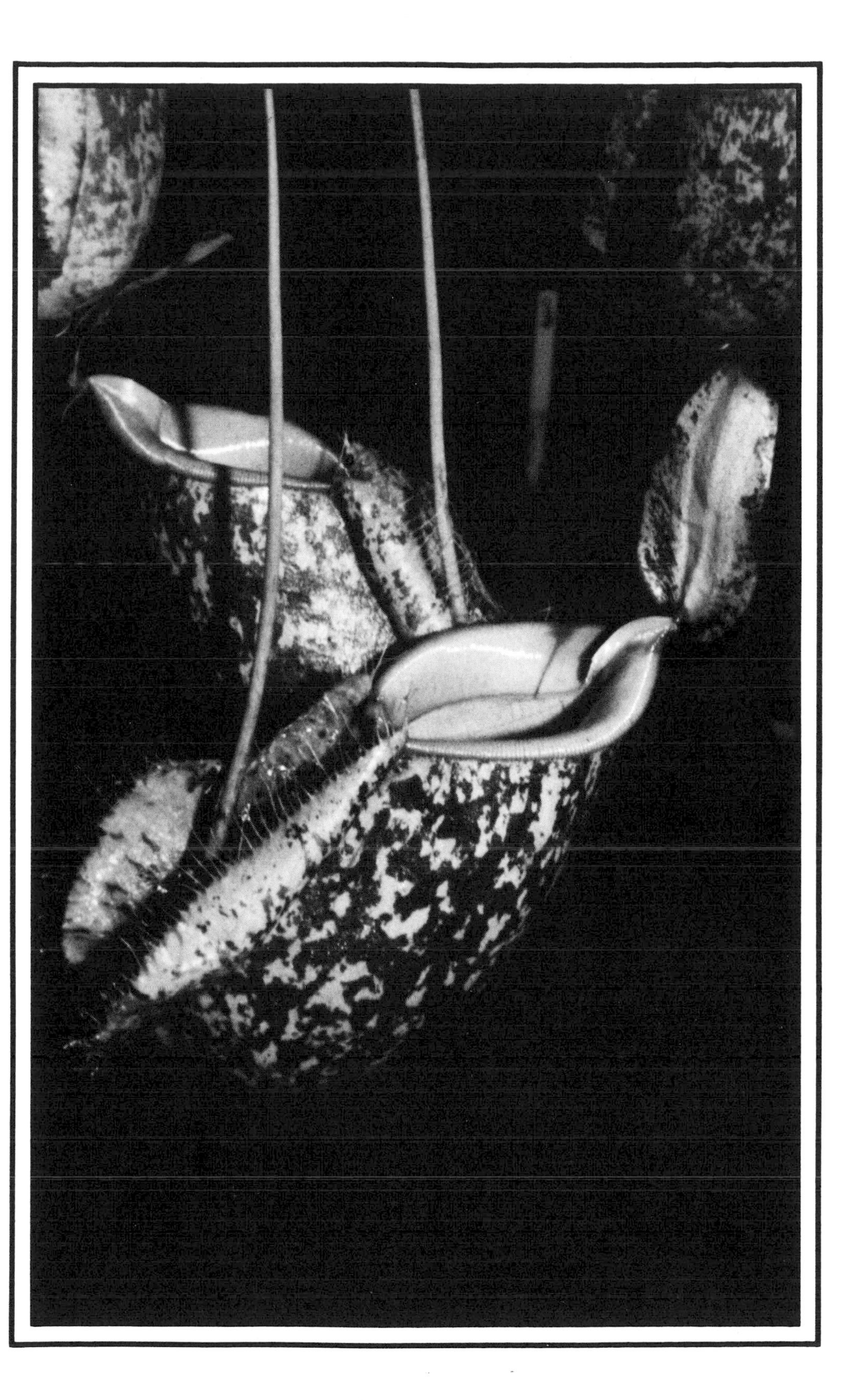

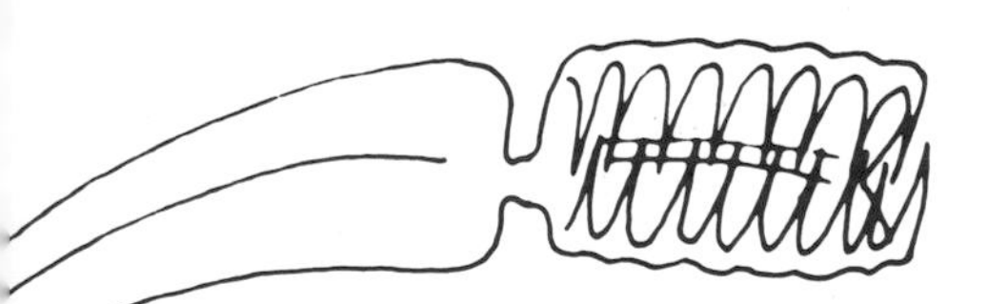

Nepenthes intermedia (commercial hybrid)

Description: Large pitchers up to six inches high. Green, mottled with red. Tall vine. The pitcher in the photograph is slightly immature, having just opened. Its hard lip has not fully formed.

Cultivation: Typical *Nepenthes* culture; available commercially. Will not withstand frost.

Reproduction: Stem cuttings, or seed with difficulty.

Nepenthes khasiana

Location: Introduced from Khasi Hills of East Hindustan, now Assam, India. Almost extinct in natural location.

Description: Tall vine. Large pitcher six inches high, long neck and flaring cap. Green, mottled with red.

Cultivation: Typical *Nepenthes* culture; difficult to obtain. Will not withstand frost.

Reproduction: Stem cuttings, or seed with difficulty.

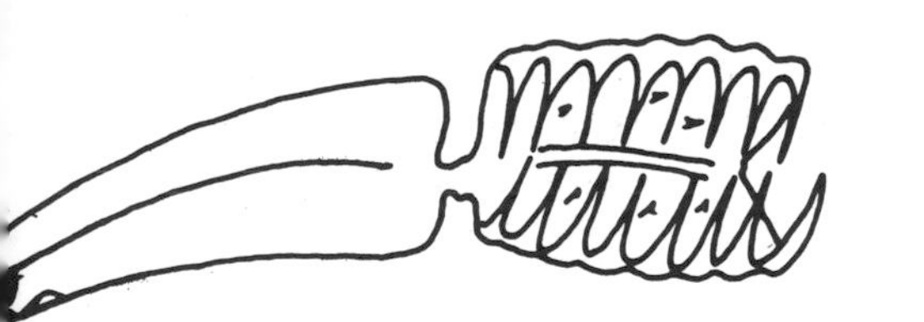

Nepenthes rafflesiania

Location: Malacca to Borneo and Sumatra.

Description: One of the most exotic looking *Nepenthes.* Large pitcher with tall neck. Lip often striped red. Pitcher is green, mottled with red. Large plants.

Cultivation: Typical *Nepenthes* culture; available commercially. Also many hybrids, often unnamed, are available from this plant. Will not withstand frost.

Reproduction: Stem cuttings, or seed with difficulty.

4. Other Carnivores

So far, we've covered about thirty species of carnivorous plants. There are actually over 450 species of carnivores. Most are slight variations of plants we've already discussed, like the 275 species of *Utricularia*, or the more than ninety species of *Drosera* that have been found.

Some, however, are very unusual, and while they're too rare or difficult to be grown at home, they really should be mentioned.

For instance, there are carnivorous fungi. These were discovered in 1888, quite by accident, by a botanist studying *Arthrobotrys obligiospora*, a species of fungus found growing around and into dead eelworms. As the botanist was peering into his microscope, a live eelworm crawled through one of the loop structures of the fungus. Instantly, the loop swelled closed, trapping the eelworm. After a violent struggle, the victim died a few

Arthrobotrys oligiospora

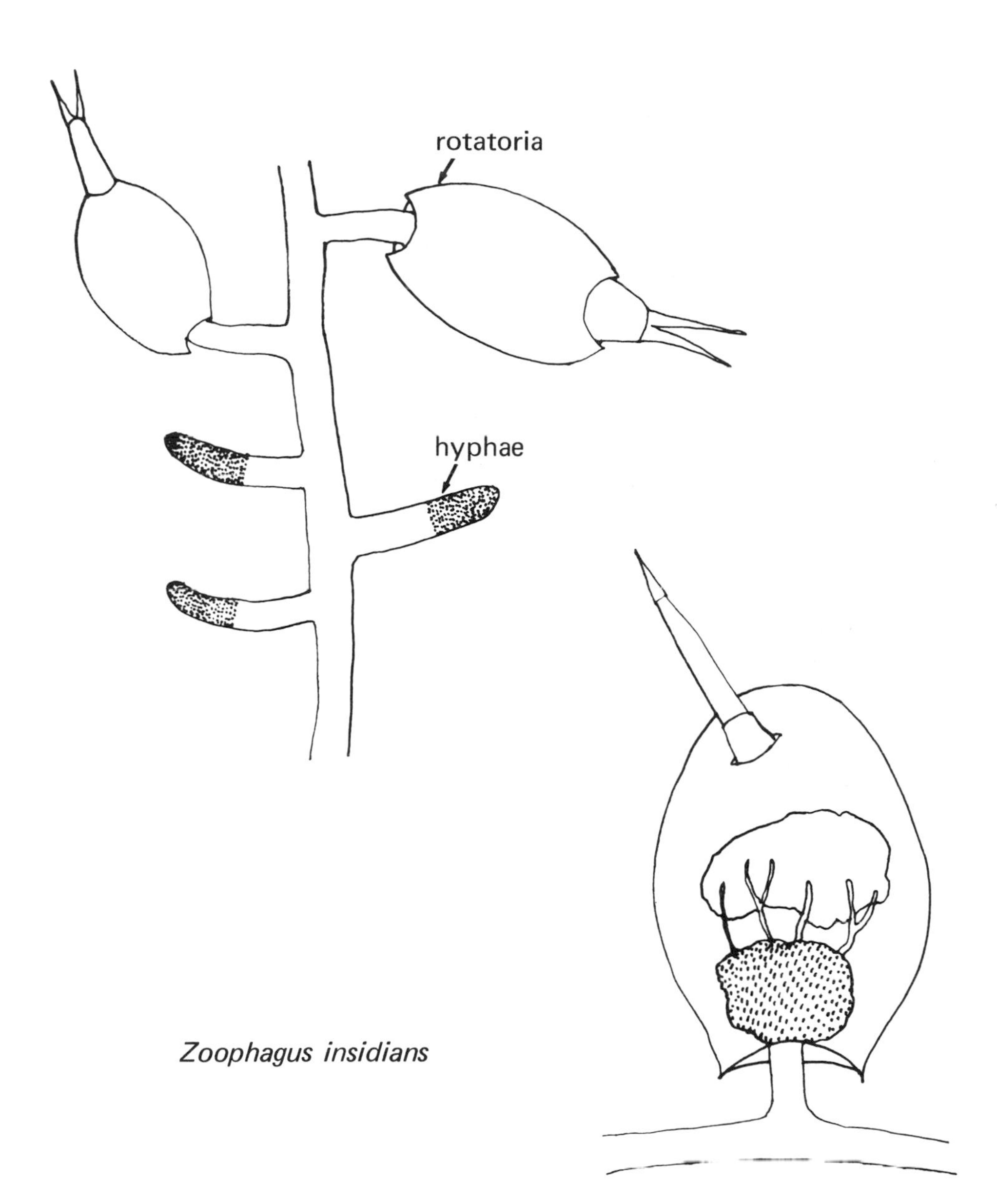

Zoophagus insidians

hours later. The fungus ingested vital nutrients from the dead animal.

The eelworm is also the victim of the *Dactyllela bebbicoides* fungus. This plant is much smaller, and the animal is many times the size of the trap. Yet, if the eelworm thrusts its head or tail into a loop, it is caught immediately.

There's one fungus that actually waits for its prey to attack first. Along the main stem of *Zoophagus insidians* are short branches called hyphae, food for a tiny animal known as rotatoria. When the animal bites into a branch, the tip swells up so large that the victim is trapped. It's very much like microscopic fishing. The animal struggles, and the plant shoots out a single glob of mucilage to help hold its prey.

The life of one species of fungus is both parasitic and carnivorous. Known as *Endocahleus asteroides*, this plant has very sticky branches. When a branch gets caught on a wandering amoeba, it breaks off from the mother plant and latches onto its host, which is so far unaffected.

Soon, however, the piece of fungus grows through the wall of the amoeba and begins absorbing nutrients from its victims. As the plant continues growing, it eventually kills the amoeba. Then the fungus forms more of the deadly branches from which it grew, and waits for another victim to come along.

Another odd carnivore is *Aldrovanda vesiculosa.* Its trap, using trigger hairs, is quite similar to the Venus flytrap, but this plant grows underwater. It can be found in fresh water from southern France to Japan, and south to Africa and Australia.

Growing about eight inches long and two inches in diameter, the plant has a very orderly arrangement. All along the stem at regular intervals are whorls with eight leaves. Each tiny trap is about three eighths of an inch long.

There are more aquatic species of carnivores. Two, *Biovularia* and *Polypompholyx*, are both so similar to *Utricularia* that only trained botanists can tell them apart.

Another aquatic carnivore is *Genlisea*, which lives submerged but blossoms above water. The plant has flat leaves and a dangling apparatus like a lobster trap. Insects entering the strange spiral shape can only go one way – further into the plant, where they are digested. The victims are tiny, since the entire plant is less than an inch long.

These aren't your garden variety carnivores, and not the kind you're likely to grow. They're quite difficult to find, but fun to know about.

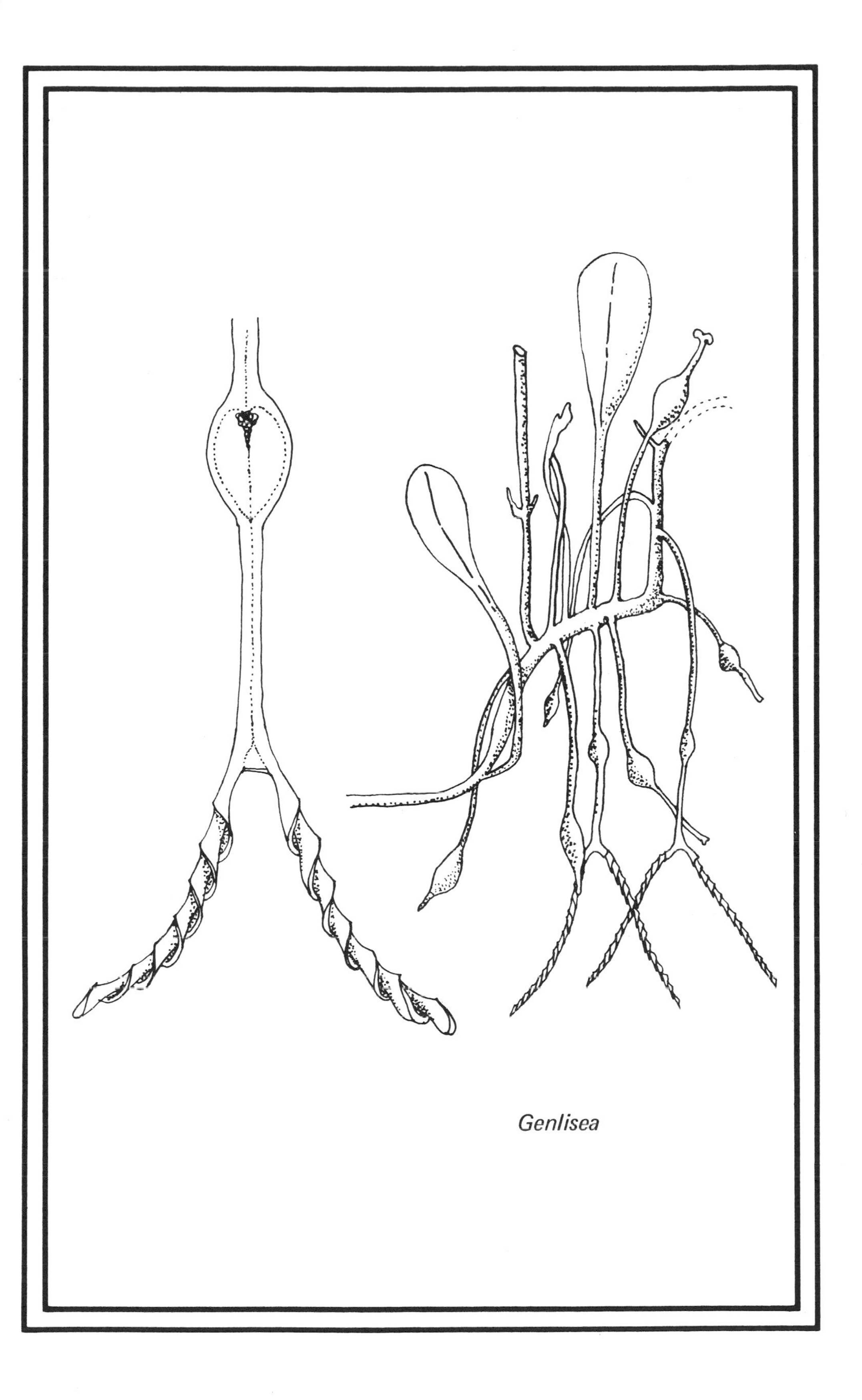
Genlisea

5. How to Grow Carnivorous Plants

Even if you don't live in a jungle, you can grow carnivorous plants at home. They aren't much more trouble than ordinary house plants, and you get the thrill of bloodcurdling adventure besides!

Feeding your plants is simple. There are probably enough insects around your house to keep them happy, if you let the plants function as natural insecticides. Sundews will eat mosquitoes and most other small insects; so will butterworts. Flytraps will take care of ants and flies, and pitcher plants devour not only mosquitoes, but roaches as well!

While carnivores can live without victims, they'll be weak, rather than thriving, healthy plants. If they're really starving, try dropping a tiny piece of hamburger, or cooked egg white, into the traps. But be careful of overfeeding, or the traps may die prematurely. As a rule, I never feed any plant more than once every six weeks, no matter how many traps it has.

Most carnivores have basically the same needs. Besides insects, they like bright light, a lot of water, high humidity, and an acid soil or moss. If you want to live with a carnivore, you'll need just a little bit of planning to produce these conditions at home. Let's take them one at a time.

Keep your plants near a window, in direct light. Northern light is usually too diffuse, but a south, east, or west window will suffice, as long as the sunlight isn't blocked by trees or other buildings. If you have a greenhouse, full sun to thirty per cent shade is fine.

Don't worry if your house seems too dark. Most carnivores grow well under artificial plant lights, such as Grow-Lux. For maximum exposure, keep the plants a few inches away from the lights. Venus flytraps should be as close as possible, and even then they may need sunlight as well. It's best not to light the plants constantly, so turn off the lights for several hours each night. Most growers find it easier to use an inexpensive timer to light the plants eighteen hours in summer, fifteen in winter, for the proper photoperiod.

Carnivorous plants grow well in an acid medium. The simplest, and also the best, is sphagnum moss, which is available in both live and dried form. Live moss, while a bit hard to come by, is preferable since it most closely resembles the plants' natural habitat. They tend to grow better, and need less repotting since live material doesn't break down. For help in getting the live moss, check the sources listed in the back of the book. Dried moss will work, too, although it must be periodically replaced. It's easily available in most plant and garden stores.

If you're using dried sphagnum, make sure it's the whole, unmilled moss, not the dustlike variety. Let it soak in water a few hours before you use it. Some growers mix sand or pearlite or soil into the moss, but I just use plain sphagnum.

Almost all the carnivores grow naturally in bogs, and they like the same conditions at home. Grow them very wet, never allowing them to dry out. One easy way to do this is to put your plants, pot and all, in a large saucer or tray full of water. Or plant them in a clear, shallow box with no drainage, but gravel at the bottom. See that there's always water at the gravel layer. If you're using a normal drainage pot, at least daily watering is necessary.

Many growers use distilled water, but it usually isn't necessary. It really depends upon the condition of your local water. Watch for mineral salt build-up in the moss. It looks like white frosting on the tips of the sphagnum. To wash out the salts, put the pot under slowly running water for a while, whenever it seems to need it. In hard water areas, you may need distilled water if your plants are in trays or non-draining containers. Remember, though, that plants without drainage require less water. One jug of distilled water will go a long way. Or, you can use rainwater or melted snow as an alternative.

While most carnivores adhere to the wetness rule, there are some exceptions. The few plants that like a drier environment, like *Drosophyllum*, the tuberous *Drosera*, and a very few *Nepenthes*, are very rare and usually available only to experienced growers.

Carnivores need moisture not only in the pot, but in the air; in other words, humidity. The same tray of water that is keeping your plants wet can help here too. If you use a tray much larger than the pot, evaporation will help raise the humidity around the plants. Misting is also a good method.

Remember that temperature affects humidity. With the same amount of water vapor in the air, if the temperature rises, the humidity is lowered. Air conditioners also pull moisture out of the air. If your plants are subjected to very dry air, watch for signs of

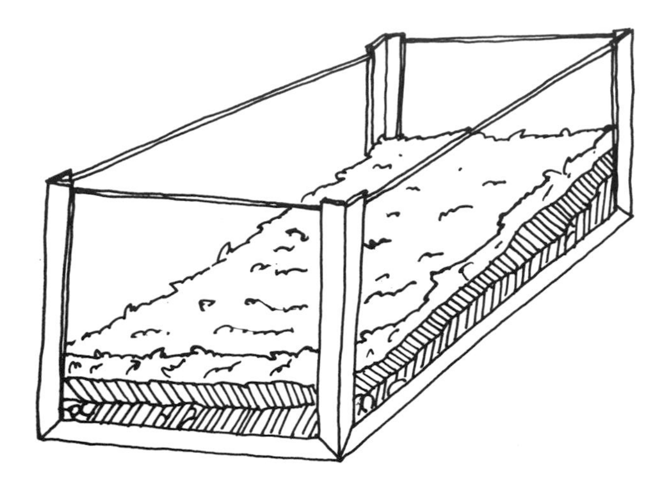

low humidity – sundews will look dry, developing pitchers will wilt, and flytraps won't close or develop new traps.

You might find your best bet for a controlled environment for your plants is a terrarium. The enclosed container helps to retain vital moisture while letting in sunlight.

Many people grow small plants in oversized brandy snifters, but any clear container with high or inward sloping sides will work perfectly. My own quick dime store survey yielded a half-dozen suitable containers. If the air in your home is particularly dry, you might want to stretch plastic wrap across the top of the terrarium to seal it. Watch out for heat building up in direct sunlight, though, and open the terrarium a bit if necessary.

For a larger container, you could convert an aquarium, sliding a piece of glass across the top to cut down evaporation. Clear plastic sweater or shoe boxes with lids also work well. And you can find many well-designed terrariums in plant and department stores.

Once you've found the terrarium you want, here's how to plant it. First, lay down a shallow layer of pearlite or gravel on the bottom of the container, for drainage. Next, an inch or two of acid soil, if you can get it, or use peat. If you make sure not to use a sterilized soil or moss, then you'll be likely to get plenty of insect eggs ready to develop and provide your plants with a steady diet. Now put down a couple of inches of sphagnum moss, and you're ready for the plants.

If your terrarium is large enough, you can slope the moss to give you a shallow, wet spot and a higher, slightly drier area. *Sarracenia* will like the wettest spot the best. *Utricularia*, *Nepenthes*, and Venus flytrap can be planted in the middle level, with *Drosera* and *Pinguicula* taking the high point.

As you increase your collection, the next serious step is a window greenhouse. It's a fine way to have a greenhouse if you can't build a regular one.

Basically, the structure is a wooden frame with glass or lucite shelves, placed in your window. Paint the wood white to reflect the maximum amount of light. When you have the glass cut to size, be sure to have the edges buffed to prevent cuts. Skip a shelf or two for taller plants.

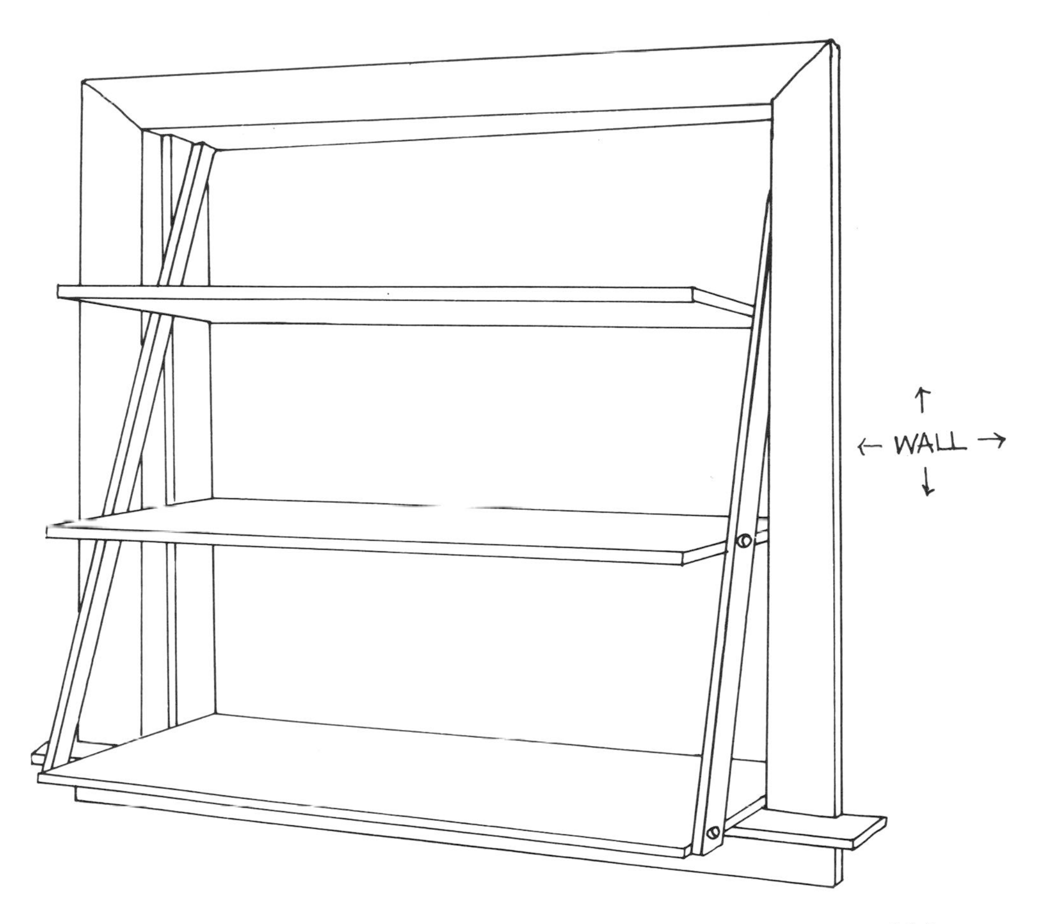

The top, sides, and front of the greenhouse should be covered with plastic. Plexiglass is good, with holes drilled so it can be screwed in place. Be sure to hinge the front for easy access. Or you could staple on a roll of polyethylene or even plastic wrap, letting it hang down so you can get in under it. A more sophisticated unit is available from Lord & Burnham of Irvington, New York.

If you have room outdoors, you can grow the plants outside if your weather is humid enough. Some plants can even survive freezing, so you may be able to keep them out during the winter. It depends upon the individual plant and your local climate, so select carefully.

To keep them outdoors, put the plants in clay pots filled with sphagnum. You can bury them in the ground, pots and all, and they'll dry out more slowly. Remember, they like the soil wet. If you have any water like a pond on your land, and the soil nearby is damp enough, you can have permanent plantings.

Carnivores, like most other plants, go through a period of dormancy in the winter. Usually the plant remains intact, but stops growing. When the plant detects cooler night temperatures and shorter days, it begins to enter the dormancy stage. Allow for these signals if your plants are in an enclosed container under lights. Dormancy is natural and healthy for the plants, and this phase should not be disturbed.

Remember, a pot, terrarium or greenhouse is only a tool for growing plants. Don't get so involved with the tools that you forget the plants.

What I've learned is from my own experience; yours may be different. There are no hard-and-fast rules.

The ideal is to have plants that grow larger each year, blossom, and thrive. If each season brings a smaller and weaker plant, something is wrong. Vary something: more light, perhaps; maybe more water. Experiment until your plants are happy. It's well worth the extra work.

I'm sure that you'll enjoy these green growing pets. Good luck.

6. Science Fiction and Mythology

THE SNAKE-TREE

The "Snake-tree" is described in a newspaper paragraph as found on an outlying spur of the Sierra Madre, in Mexico. It has sensitive branches of a slimy, snaky appearance, and when a bird alights on them incautiously, it is seized, drawn down in the tree and lost to sight. Soon afterward it falls, flattened out, to the ground, where bones and feathers, no doubt of former captures, cover the earth. An adventurous traveler, having touched one of the branches of the tree, tells how it closed up on his hand with such force that it tore the skin when he wrenched it away. He then fed the tree with chickens, and the tree absorbed their blood by means of suckers with which its branches were covered, very much like those of the octopus.

THE MAN-EATING TREE

Dr. Carle Liche, who claimed to have seen the man-eating tree in Madagascar in 1878, was quoted in a newspaper story as follows:

"It was while among these natives that I was witness to what was probably the most horrible sight I have ever seen. Their religion consists in the worship of their sacred tree, to which they offer human sacrifice.

"One evening I was welcomed to witness their ceremony. I followed the tribe as they made their way to the sacred tree.

"It was ten feet high, with eight leaves, ten to twelve feet long, plentifully strewn with huge venomous looking thorns. From underneath there grew half a dozen frail looking stamen.

"The natives, suddenly springing up, rushed upon one poor girl with unearthly shrieks and yells. They surrounded her, and with shouts and gestures ordered her to climb the tree. Terrified she shrank back, apparently begging for mercy.

"At last, seeing it was useless, she turned and faced the tree. For a moment she stood still, gathering herself up for a supreme effort, then she scrambled up, and reaching the top knelt and drank of the holy liquid.

"Suddenly I realized what was happening, and I seemed to be paralyzed with horror. The tree, seemingly so dead and motionless a moment before, had come to life. The stamen, so frail looking, had coiled themselves about the girl's head and shoulders, holding her so firmly that all her efforts to free herself remained absolutely useless.

"The green branches began to writhe, and coiled themselves round and round like snakes. Then as she struggled, the great leaves began to rise slowly, very slowly. Those evil looking thorns were closing in on her with the force of a hydraulic press.

"As they came together tightly there trickled down the trunk a pinkish mixture. The maddened natives fought each other to get one mouthful of the intoxicating fluid from the tree and the blood of the human sacrifice.

"Then feasting began amid much rejoicing. The devil was appeased."

Sacrificed to a man-eating plant. *American Weekly,* September 26, 1920.

A FLESH-EATING VINE

Mr. Dunstan, a naturalist, relates that while botanizing in the swamps of Nicaragua hunting for specimens, he heard his dog crying out, as if in agony. Running to the spot whence the cries came, he found the animal caught in a plant composed entirely of bare interlacing stems, resembling the branches of a weeping willow denuded of foliage, but of a dark, nearly black, color, and covered with a thick gum that exuded from its surface. Mr. Dunstan attempted to cut the poor beast free with his knife, but the plant's twigs curled like living, sinuous fingers about Mr. Dunstan's hand. It required great force to free it from their grasp, which left the flesh red and blistered. The tree, it is reported, is well known to the natives, who tell many stories of its death-dealing powers. Its appetite is voracious and insatiable; and in a few minutes it will suck the nourishment from a large lump of meat, rejecting the carcass as a spider does that of a used fly.

ANOTHER MAN-EATING TREE

" . . . The whole thing had changed shape and was horribly alive and alert. The dull, heavy leaves had sprung from their compact formation and were coming at him from all directions, advancing on the ends of long vine-like stems which stretched across like the necks of innumerable geese and, now that the old man had stopped his screaming, the air was full of hissing sounds.

"The leaves did not move straight at their target, but with a graceful, side-to-side sway, like a cobra about to strike. From the far side, the distant leaves were peeping and swaying on their journey around the trunk and even the tree top was bending down to join in the attack. The bending of the trunk was spasmodic and accompanied by sharp cracks.

"The effect of this advancing and swaying mass of green objects was hypnotic, like the charm movements of a snake. Bryant could not move, though the nearest leaf was within an inch of his face. He could see that it was armed with sharp spines on which a liquid was forming. He saw the heavy leaf curve like a green-mittened hand, and as it brushed his eyebrows in passing he got the smell of it – the same animal smell that hung in the surrounding air. Another instant and the thing would have had his eyes in its sticky, prickly grasp, but either his weakness or the brown man's strength threw them both on their backs.

Escaped from the embrace of the man-eating tree. *American Weekly,* January 4, 1925.

"The charm was broken. They crawled out of the circle of death and lay panting in the grass while the malignant plant, cracking and hissing, yearned and stretched and thrashed to get at them. . . ."

Suddenly a leaf bent lower than usual and touched the Patchwork Girl. Swiftly it enveloped her in its embrace, covering her completely in its thick folds, and then it swayed back upon its stem.

"Why, she's gone!", gasped Ojo, in amazement, and listening carefully he thought he could hear the muffled screams of Scraps coming from the center of the folded leaf. But, before he could think what he ought to do to save her, another leaf bent down and captured the Glass Cat, rolling around the little creature until she was completely hidden, and then straightening up again upon its stem.

– from *The Patchwork Girl of Oz* by L. Frank Baum, copyright 1913

THE DEATH FLOWER

"In 1581, the valiant explorer Captain Arkright learned of an atoll in the South Pacific that one might not visit, save on peril of his life, for this coral ring enclosed a group of islets on one of which the Death Flower grew; hence it was named El Banoor, or Island of Death. This flower was so large that a man might enter it – a cave of color and perfume – but if he did so it was the last of him, for, lulled by its strange fragrance, he reclined on its lower petals and fell into the sleep from which there is no waking. Then as if to guard his slumber, the flower slowly folded its petals about him. The fragrance increased and burning acid was distilled from its calyx, but of all hurt the victim was unconscious. and so passing into death through splendid dreams, he gave his body to the plant for food."

MONKEY-TRAP TREE

A recent report is credited to a Brazilian explorer named Mariano da Silva who returned from an expedition that led him into a district of Brazil that borders on Guyana. He had there sought out the settlement of Yatapu Indians. During his journey he saw a tree which nourishes itself on animals. The tree itself exudes a peculiar sharp odor which attracts its victims, especially monkeys. As soon as they climb the trunk, all is up with them, for very quickly they are completely closed in by the leaves, and one neither hears nor sees them again. After about three days the leaves open and let drop to the earth the bones, completely stripped.

Where to Obtain Carnivorous Plants

A note to the buyer: all *Dionaeae*, *Darlingtoniae*, and most *Sarraceniae* are field-collected plants. Others are usually nursery grown.

Armstrong Associates, Inc.
P. O. Box 127
Basking Ridge, New Jersey 07920

A mail-order operation specializing in selling large quantities of *Dionaeae* to schools and institutions. Operated by Alan Swenson, who is also involved in other areas of gardening. His mail order service is quick and efficient, although some plants may have been subjected to cold storage, which confuses seasons for the plants. Also available: American *Drosera*, *Sarraceniae*, *Pinguiculae*, and *Darlingtonia*; sometimes, *Nepenthes* and non-American *Drosera.* Catalogue 25 cents.

Insectivorous Botanical Gardens
1918 Market Street
P. O. Box 3322
Wilmington, North Carolina 28403

Mrs. Northrop and her family supply virtually all the *Dionaeae* sold throughout the world. These gardens have done more to promote the popularity of carnivorous plants than any other single group. All their plants are field-grown. Also available: *Sarraceniae*, *Drosera rotundifolia, Darlingtonia,* and *Pinguiculae.* This is the oldest and best nursery for carnivores. Catalogue 25 cents.

King's Park and Botanic Garden
Perth, Western Australia 6005

Seed of Australian species is often available upon request – in particular, *Drosera*, *Byblis*, and *Cephalotus.*

Marcel Lecoufle
5, rue de Paris
94470 Boissy St.-Leger
France

Lecoufle is one of the finest names associated with horticulture in Europe. His family has a long established and truly great nursery located on beautiful grounds outside Paris. They are generally not open to the public.

His plants are well grown and his collection of carnivores is extensive. Long experience in packing and handling plants virtually assures safe delivery anywhere if shipped by air. His stock includes *Dionaeae*, *Drosera*, *Pinguiculae*, and *Nepenthes*. Write for a price and availability list. He periodically publishes a fine color-illustrated catalogue.

Randall Schwartz
P. O. Box 283
Lenox Hill Station
New York, New York 10021

The author. Every plant in this book is available periodically. Write for an availability/price list.

The following are firms I do not know personally, but they are reputable:

Antonelli Brothers
2545 Capitola Road
Santa Cruz, California 95062

Baywood Nurseries
P. O. Box 24
Plymouth, Florida 32768

G. Ghose and Company
Townend
Darjeeling, India
Sells only the native *Nepenthes khasiana.*

Plant Exchanges, Services, and Societies

One of the most common ways of enlarging a plant collection is that of swapping with other growers, normally a very haphazard arrangement involving great difficulties and much wasted correspondence.

All of this has changed, however, for there exists for carnivorous plant growers a computerized plant exchange service. You merely send in a list of carnivores that you are currently growing along with the names of species you desire, and you will be matched to someone who has the plants you want and perhaps wants *your* plants. You then correspond directly with the grower.

This is actually more than just a plant swap service. Bob Zeimer, who runs it with no finances and much dedication, is very interested in disseminating carnivorous plant material as widely as possible. Often, seed or cuttings will be furnished without calling for plants in exchange. It is imperative, however, to send a list of the carnivores you're growing and how you are growing them in order for them to determine your abilities. They have no desire to waste extremely difficult or rare plant material on someone who is ready only for fairly easy plants.

There is also a *Carnivorous Plant Newsletter* published by David Schnell and Joseph Mazrimas which appears quarterly. This is a good bulletin, often illustrated. It includes articles on culture, taxonomy, and research and provides a central clearinghouse for members. It makes it quite easy to find people in your area who grow carnivores.

I highly recommend getting a subscription and at $2.00 a year it is truly a bargain. Outside of North America the rate is $3.00 per year. Write to either:

D. E. Schnell
Rt. 4, Box 275B
Statesville, North Carolina 28677

J. A. Mazrimas
329 Helen Way
Livermore, California 94550

There is a rather extensive Insectivorous Plant Society of Japan which has been indirectly responsible for much hybridizing of carnivorous plants. Their bulletin, published since 1950, is printed only in Japanese, but the plants are identified by their Latin names. The address is:

Mr. S. Komiya
Department of Biology
Nippon Dental College
9-20-1-Chome, Fujimi
Chiyoda-ku, Tokyo, Japan

Botanical Gardens Where Carnivorous Plants Can Be Seen

Since the popularity of carnivorous plants is so fast growing, check with any botanical garden, even if not listed here.

U.S.A.

Brooklyn Botanical Gardens, Brooklyn, New York.

California State University, Fullerton, California

California State University, Humboldt, California

Columbia Zoological Park and Botanical Garden, Riverbanks Park Commission, Columbia, South Carolina.

Longwood Botanical Gardens, Kennett Square, Pennsylvania. A particularly complete collection of *Nepenthes*.

Los Angeles State and County Arboretum, Arcadia, California.

Missouri Botanical Gardens, 2315 Tower Grove Avenue, St. Louis, Missouri

New York Botanical Garden, Bronx, New York

Phipps Conservatory, Schenley Park, Pittsburgh, Pennsylvania

Sán Francisco Conservatory, Golden Gate Park, San Francisco, California

University of California Botanical Gardens, Berkeley, California. A particularly complete collection.

University of North Carolina at Chapel Hill, North Carolina

BRITAIN

Royal Botanic Garden, Edinburgh 3, Scotland

Royal Botanic Gardens, Kew, Richmond, Surrey. A particularly complete collection.

University Botanic Garden, Cambridge

University Botanic Garden, Oxford

CANADA

Montreal Botanic Garden, 4101 Sherbrooke Street, Montreal, Quebec

GERMANY

The Munich Botanic Garden, Munich. A particularly complete collection.

IRELAND

National Botanic Gardens, Glasnevin

NEW ZEALAND

The Botanic Garden, Auckland

Suggested Reading

Erikson, Rica. *Plants of Prey.* Lamb Publications Pty., Ltd., Australia, 1968.

Lloyd, Francis Ernest. *The Carnivorous Plants.* Waltham, Mass., 1942.

Darwin, Charles. *Insectivorous Plants.* New York: D. Appleton & Co., 1892.

Graf, A. B. *Exotica.* Roehrs Company, Inc., Rutherford, New Jersey.